TEST ITEM BOOK

Volume I: Review and Preview - Chapter 10

For
Stewart's CALCULUS, Second Edition

By

James Stewart

With Contributions From

Edward Spitznagel
Joan Thomas
and
Engineering Press

Brooks/Cole Publishing
Pacific Grove, California

Brooks/Cole Publishing Company
A Division of Wadsworth, Inc.

Printed in the United States of America

5   4   3   2   1

ISBN 0-534-13218-9

CONTENTS

Calculus, 2nd Edition
by James Stewart
Review and Preview, Section 1
Numbers, Inequalities, and Absolute Values

1. The solution set of the inequality $|x - 1| < 12$ is of the form $(a, 13)$. Find the value of a.

    A) -12 B) -9 C) -11 D) -6

    E) -7 F) -13 G) -10 H) -8

    Answer: -11 easy

2. The solution set of the inequality $|2 - x| \geq 10$ is of the form $(-\infty, -8] \cup [a, \infty)$. Find the value of a.

    A) -7 B) 11 C) -4 D) 8

    E) 12 F) -6 G) 9 H) 10

    Answer: 12 easy

3. The solution set of the inequality $|3x - 2| \geq 7$ is of the form $(-\infty, a] \cup [3, \infty)$. Find the value of a.

    A) $-7/5$ B) $-7/2$ C) $-5/7$ D) $-5/2$

    E) $-2/5$ F) $-5/3$ G) $-2/7$ H) $-7/3$

    Answer: $-5/3$ medium

4. The solution set of the inequality $x^2 - 2x \leq 15$ is of the form $[a, 5]$. Find the value of a.

    A) 2 B) -4 C) 0 D) -3

    E) -5 F) -1 G) -2 H) 1

    Answer: -3 medium

5. The solution set of the inequality $x^2 + x - 1 \geq 0$ is an interval of the form $[a, b]$. Find the value of $b - a$ (that is, the length of the interval).

    A) $\sqrt{2}$ B) 3 C) 2 D) $3/2$

    E) $5/2$ F) $\sqrt{3}/2$ G) $\sqrt{3}$ H) $\sqrt{5}$

    Answer: $\sqrt{5}$ medium

6. The solution set of the inequality $|x| + |x-1| \leq 5$ is an interval of the form $[a, b]$. Find the value of $b - a$ (that is, the length of the interval).

A) 6   B) 7/5   C) 3/2   D) 7/3

E) 3   F) 5/2   G) 4   H) 5

Answer: 5 hard
_____

7. Find the larger of the two roots of the equation $\left|\dfrac{2x-1}{x+1}\right| = 3$.

A) $-4/3$ B) $-3/4$ C) $-2/3$ D) $-5/4$

E) $-3/5$ F) $-5/2$ G) $-2/5$ H) $-3/2$

Answer: $-2/5$ hard
_____

8. Given $|x-4| = 2x + 1$ find the value of x.

A) -1 B) -5 C) 5   D) 1

E) 0   F) 2   G) -2 H) 3

Answer: 1 medium
_____

Calculus, 2nd Edition
by James Stewart
Review and Preview, Section 2
Coordinate Geometry and Lines

1. Find the slope of the line through the two points $(1, 3)$ and $(7, 13)$.

   A) 5/3   B) −4/7   C) −7/4   D) 4/7

   E) −3/5   F) 7/4   G) 3/5   H) −5/3

   Answer: 5/3 easy
   _____

2. Find the y-intercept of the line through the two points $(4, 9)$ and $(18, 2)$.

   A) 9.5   B) 9   C) 12   D) 10.5

   E) 11   F) 11.5   G) 10   H) 12.5

   Answer: 11 medium
   _____

3. Find the y-intercept of the line passing through the point $(6, 5)$ and perpendicular to the line $y = 2x + 17$.

   A) 6   B) 10   C) 12   D) 8

   E) 0   F) 2   G) −2   H) 4

   Answer: 8 medium
   _____

4. Find the slope of the line that has y-intercept equal to 3.5 and passes through the point $(3, 5)$.

   A) −2/3   B) 3/2   C) −2   D) 2/3

   E) 2   F) 1/2   G) −3/2   H) −1/2

   Answer: 1/2 easy
   _____

5. Find the angle of inclination, in degrees, of the line through the two points $\left(0, \sqrt{3}\right)$ and $(1, 0)$.

   A) 45   B) 30   C) 60   D) 90

   E) 120   F) 180   G) 150   H) 135

   Answer: 120 easy
   _____

6. Find the x-coordinate of the point of intersection of the two lines $y = 2x + 13$ and $y = 4x - 49$.

A) -18 B) 18 C) -62 D) -36

E) 31 F) 36 G) -31 H) 62

Answer: 31 medium
_____

7. Find the y-intercept of the perpendicular bisector of the line segment joining the two points $(3, 2)$

and $(6, 3)$.

A) -16 B) 8 C) 9 D) 16

E) 12 F) -8 G) -12 H) -9

Answer: 16 hard
_____

8. Find the distance between the two points $(0, 7)$ and $(3, 3)$.

A) $\sqrt{4}$ B) 4 C) $\sqrt{5}$ D) $\sqrt{2}$

E) 3 F) $\sqrt{3}$ G) 5 H) 2

Answer: 5 easy
_____

9. Find the distance between the two points $(0, 1)$ and $(2, 0)$.

A) $\sqrt{4}$ B) $\sqrt{2}$ C) 3 D) 4

E) 2 F) 5 G) $\sqrt{3}$ H) $\sqrt{5}$

Answer: $\sqrt{5}$ easy
_____

10. Find the point on the x-axis that is equidistant from $(1, 1)$ and $(2, 2)$.

A) $(1.5, 0)$ B) $(-1.5, 0)$ C) $(2, 0)$ D) $(-2.5, 0)$

E) $(-3, 0)$ F) $(3, 0)$ G) $(2.5, 0)$ H) $(-2, 0)$

Answer: $(3, 0)$ medium
_____

11. Find the x-coordinate of the midpoint of the line segment joining the points $(35, 29)$ and

$(-117, 22)$.

A) $51/2$ B) -41 C) $-51/2$ D) $-7/2$

E) 76 F) -76 G) $7/2$ H) 41

Answer: -41 medium
_____

12. Find the equation of the perpendicular bisector of the segment joining the points $(-4, 0)$ and $(8, 6)$.

A) $2x + y = 7$ B) $x + y = 7$ C) $x + 2y = 7$

D) $x + y = 3$ E) $2x + y = 3$ F) $x + 2y = 3$

G) $x - 2y = 4$ H) $2x - y = 4$

Answer: $2x + y = 7$ medium

13. Given the points $A(6, -7)$, $B(-3, -1)$, and $C(2, -2)$ use slopes to show that ABC is a right triangle.

Answer: $m_{AB} = \dfrac{-3 - (-7)}{11 - 6} = \dfrac{4}{5}$ and $m_{AC} = \dfrac{-2 - (-7)}{2 - 6} = -\dfrac{5}{4}$. Thus $m_{AB} \cdot m_{AC} = -1$

and so AB is perpendicular to AC and $\triangle ABC$ must be a right triangle. medium

14. Show that $A(1, 1)$, $B(11, 3)$, $C(10, 8)$, and $D(0, 6)$ are vertices of a rectangle.

Answer: The slopes of the four sides are: $m_{AB} = \dfrac{3 - 1}{11 - 1} = \dfrac{1}{5}$, $m_{BC} = \dfrac{8 - 3}{10 - 11} = -5$,

$m_{CD} = \dfrac{6 - 8}{0 - 10} = \dfrac{1}{5}$, and $m_{DA} = \dfrac{1 - 6}{1 - 0} = -5$. Hence $AB \parallel CD$, $BC \parallel DA$,

$AB \perp BC$, $BC \perp CD$, $CD \perp DA$, and $DA \perp AB$, and so $ABCD$ is a rectangle.

medium

15. Show that the lines $3x - 5y + 19 = 0$ and $10x + 6y - 50 = 0$ are perpendicular and find their point of intersection.

Answer: $3x - 5y + 19 = 0 \Leftrightarrow 5y = 3x + 19 \Leftrightarrow y = \frac{3}{5}x + \frac{19}{5} \Rightarrow m_1 = \frac{3}{5}$ and $10x + 6y - 50 =$

$0 \Leftrightarrow 6y = -10x + 50 \Leftrightarrow y = -\frac{5}{3}x + \frac{25}{3} \Rightarrow m_2 = -\frac{5}{3}$. Since $m_1 m_2 = \frac{3}{5}\left(-\frac{5}{3}\right) = -1$ the two

lines are perpendicular. To find the point of intersection: $\frac{3}{5}x + \frac{19}{5} = -\frac{5}{3}x + \frac{25}{3} \Leftrightarrow$

$9x + 57 = -25x + 125 \Leftrightarrow 34x = 68 \Leftrightarrow x = 2 \Rightarrow y = \frac{3}{5}(2) + \frac{19}{5} = \frac{25}{5} = 5$. Thus, the point

of intersection is $(2, 5)$. medium

1. Find the y-coordinate of the vertex of the parabola $y + 3 = x^2$.

   A) $-4/9$ B) $-9/4$ C) $9/4$ D) $-3/2$

   E) $3/2$ F) $4/9$ G) 3 H) -3

   Answer: -3 easy

2. Find the x-coordinate of the vertex of the parabola $y = x^2 + x$.

   A) -4 B) $-1/4$ C) $-1/2$ D) $1/2$

   E) 4 F) -2 G) 2 H) $1/4$

   Answer: $-1/2$ medium

3. Find the positive x-intercept of the ellipse $4x^2 + 9y^2 = 36$.

   A) 3 B) 6 C) 8 D) 4

   E) 9 F) 12 G) 2 H) 8

   Answer: 3 medium

4. Find the x-coordinate of the center of the ellipse $2x^2 + 6x + y^2 = 0$.

   A) 3 B) $-3/2$ C) $3/2$ D) 2

   E) $1/2$ F) -3 G) -2 H) $-1/2$

   Answer: $-3/2$ medium

5. Find the slope of the positive-sloped asymptote of the hyperbola $\dfrac{x^2}{2^2} - \dfrac{y^2}{3^2} = 4$.

   A) $3/2$ B) $1/2$ C) $1/3$ D) $2/3$

   E) 2 F) 3 G) $4/9$ H) $9/4$

   Answer: $3/2$ medium

6. For what value of the number c is the hyperbola $4x^2 - cy^2 = 144$ an equilateral hyperbola?

A) 12   B) 72   C) 1   D) 6

E) 9   F) 36   G) 4   H) 576

Answer: 4 medium _____

7. Find the x-coordinate of the point or points farthest to the right in the region bounded by the curves $y = 3x$, $x = y^2$.

A) 9   B) $\sqrt{3}/3$ C) 1/27   D) 1/9

E) $\sqrt{3}$ F) 3   G) 1/3   H) 27

Answer: 1/9 medium _____

8. Find the radius of the circle $x^2 + y^2 - 2x + 4y = 4$.

A) $\sqrt{3}$ B) 4   C) 3   D) $\sqrt{5}$

E) 5   F) 2   G) $\sqrt{2}$ H) 1

Answer: 3 medium _____

9. Find the x-coordinate of the center of the circle $x^2 + 3x + y^2 + 6y = 15$.

A) 5/2   B) 1/2   C) $-1/2$ D) $-5/2$

E) 3/2   F) -3 G) $-3/2$ H) 3

Answer: $-3/2$ medium _____

10. For what value of c does the equation $x^2 + 2x + y^2 + 4y = c$ represent not a circle but rather a single point?

A) 6   B) -4 C) -6 D) 5

E) 4   F) -3 G) 3   H) -5

Answer: -5 medium _____

Calculus, 2nd Edition
by James Stewart
Review and Preview, Section 4
Functions and Their Graphs

1. Find the largest value in the domain of the function $f(x) = 7 - 2x$, $-1 \le x \le 6$.

    A) 9   B) 6   C) 1   D) 2

    E) −1   F) −7   G) 7   H) −2

    Answer: 6 easy

2. Find the largest value in the range of the function $f(x) = 7 - 2x$, $-1 \le x \le 6$.

    A)7   B)1   C)9   D)−7

    E)−2   F)2   G)6   H)−1

    Answer: 9 easy

3. Find the smallest value in the domain of the function $f(x) = \sqrt{2x - 5}$.

    A)2   B)5/2   C)5   D)2/5

    E)−2   F)1   G)0   H)−5

    Answer: 5/2 easy

4. Find the largest value in the range of the function $f(x) = 3 - 2x^2$.

    A)−2/3   B)2/3   C)3   D)−2

    E)−3   F)−3/2   G)2   H)3/2

    Answer: 3 medium

5. The domain of the function $f(x) = \sqrt{1 - x - x^2}$ is a closed interval $[a, b]$. Find its length $b - a$.

    A)$\sqrt{10}/2$ B)$\sqrt{2}/2$ C)$\sqrt{3}$ D)$\sqrt{5}$

    E)$\sqrt{5}/2$ F)$\sqrt{10}/3$ G)$\sqrt{2}$ H)$\sqrt{3}/2$

    Answer: $\sqrt{5}$ medium

6. Find the smallest value in the range of the function $f(x) = |2x| + |2x + 3|$.

A)2   B)3   C)5   D)1/2

E)3/2   F)5/2   G)0   H)1

Answer: 3 medium

7. Find the smallest value in the domain of the function $f(x) = \sqrt{\frac{x}{\pi - x}}$.

A)0   B)$\pi - 2$ C)$\pi/2$ D)$\pi - 1$

E)$-\pi$ F)$1 - \pi$ G)$-\pi/2$ H)π

Answer: 0 medium

8. A rectangle has an area of 16 m^2. Express the perimeter of the rectangle as a function of the length of one of its sides.

Answer: Let the length and width of the rectangle be l and w respectively. Then the area is $lw = 16$, so that $w = 16/l$. The perimeter is $P = 2l + 2w$, so $P(l) = 2l + 2(16/l) = 2l + 32/l$, and the domain of P is $l > 0$, since lengths must be positive quantities.

medium

9. Express the surface area of a cube as a function of its volume.

Answer: Let the volume of the cube be V and the length of an edge be l. Then $V = l^3$ so $l = \sqrt[3]{V}$, and the surface area will be $S(V) = 6(\sqrt[3]{V})^2 = 6V^{2/3}$, with domain $V > 0$.

Calculus, 2nd Edition
by James Stewart
Review and Preview, Section 5
Combinations of Functions

1. Let $f(x) = \sqrt{x^2 + 9}$ and let $g(x) = \sqrt{x}$. Find the value of $(f + g)(x)$ when $x = 4$.

   A) 8   B) 6   C) 2   D) 5
   E) 9   F) 7   G) 3   H) 4

   Answer: 7 easy
   _____

2. Let $f(x) = 2x + 3$ and let $g(x) = 2 - 3x$. Find the value of $(fg)(x)$ when $x = 1$.

   A) 1   B) 7   C) -1 D) 5
   E) -5 F) 3   G) 9   H) -3

   Answer: -5 easy
   _____

3. Let $f(x) = x + 3$ and let $g(x) = 2x$. Find the value of $(f \circ g)(x)$ when $x = 1$.

   A) 9   B) -5 C) -1 D) 5
   E) 7   F) 3   G) -3 H) 1

   Answer: 5 easy
   _____

4. Let $f(x) = x^2$ and let $g(x) = 2x$. Find the value of $(g \circ f)(x)$ when $x = 3$.

   A) 36   B) 144   C) 48   D) 72
   E) 18   F) 4   G) 9   H) 6

   Answer: 18 medium
   _____

5. Let $f(x) = x^2 - 1$. Find the value of $(f \circ f \circ f)(x)$ when $x = 2$.

   A) 45   B) 65   C) 27   D) 63
   E) 8   F) 91   G) 21   H) 3

   Answer: 63 medium
   _____

6. Let $f(x) = 2x$ and let $(f \circ g)(x) = x^2$. Find the value of $g(4)$.

A) 6   B) 2   C) 16   D) 8

E) 0   F) 1   G) 12   H) 4

Answer: 8   medium

7. Let $f(x) = 2x$ and let $(g \circ f)(x) = x^2$. Find the value of $g(4)$.

A) 4   B) 2   C) 16   D) 12

E) 8   F) 1   G) 0   H) 6

Answer: 4   medium

Calculus, 2nd Edition
by James Stewart
Review and Preview, Section 6
Types of Functions; Shifting and Scaling

1. Relative to the graph of $y = x^2$, the graph of $y = (x-2)^2$ is changed in what way?

    A) shifted 2 units up

    B) compressed vertically by the factor 2

    C) compressed horizontally by the factor 2

    D) shifted 2 units to the left

    E) shifted 2 units to the right

    F) shifted 2 units down

    G) stretched vertically by the factor 2

    H) stretched horizontally by the factor 2

    Answer: shifted 2 units to the right (easy)

2. Relative to the graph of $y = x^2$, the graph of $y = x^2 - 2$ is changed in what way?

    A) shifted 2 units down

    B) stretched horizontally by the factor 2

    C) shifted 2 units to the right

    D) stretched vertically by the factor 2

    E) compressed horizontally by the factor 2

    F) compressed vertically by the factor 2

    G) shifted 2 units up

    H) shifted 2 units to the left

    Answer: shifted 2 units down (easy)

3. Relative to the graph of $y = x^3$, the graph of $y = 2x^3$ is changed in what way?

    A) compressed horizontally by the factor 2   B) shifted 2 units down

    C) stretched vertically by the factor 2   D) stretched horizontally by the factor 2

    E) shifted 2 units up   F) compressed vertically by the factor 2

    G) shifted 2 units to the right   H) shifted 2 units to the left

    Answer: stretched vertically by the factor 2 (easy)

4.   Relative to the graph of $y = x^2 + 2$, the graph of $y = 4x^2 + 2$ is changed in what way?

A) compressed vertically by the factor 2
B) stretched horizontally by the factor 2
C) compressed horizontally by the factor 2
D) shifted 2 units up
E) shifted 2 units to the right
F) stretched vertically by the factor 2
G) shifted 2 units to the left
H) shifted 2 units down

Answer: compressed horizontally by the factor 2 (medium)

5.   Relative to the graph of $y = \sin x$, the graph of $y = \sin(x - 3)$ is changed in what way?

A) compressed horizontally by the factor 3
B) shifted 3 units to the right
C) compressed vertically by the factor 3
D) shifted 3 units up
E) shifted 3 units to the left
F) stretched vertically by the factor 3
G) shifted 3 units down
H) stretched horizontally by the factor 3

Answer: shifted 3 units to the right (medium)

6.   Relative to the graph of $y = x^2$, the graph of $y = 4x^2$ can be thought of as expanded vertically by the factor 4. In what other way can it be thought of as being changed?

A) shifted 2 units down
B) shifted 4 units down
C) stretched horizontally by the factor 2
D) compressed horizontally by the factor 4
E) shifted 2 units up
F) shifted 4 units up
G) compressed horizontally by the factor 2
H) stretched horizontally by the factor 4

Answer: compressed horizontally by the factor 2 (medium)

7.   Relative to the graph of $y = \sin x$, where x is in radians, the graph of $y = \sin x$, where x is in degrees, is changed in what way?

A) stretched vertically by the factor $180/\pi$
B) compressed horizontally by the factor $180/\pi$
C) compressed horizontally by the factor $90/\pi$
D) stretched horizontally by the factor $90/\pi$
E) compressed vertically by the factor $90/\pi$
F) stretched vertically by the factor $90/\pi$
G) stretched horizontally by the factor $180/\pi$
H) compressed vertically by the factor $180/\pi$

Answer: stretched horizontally by the factor $180/\pi$ (hard)

Calculus, 2nd Edition
by James Stewart
Chapter 1, Section 1
The Tangent and Velocity Problems

1. For the curve $y = x^2$, find the slope of the tangent line at the point $(3, 9)$.

    A) 2   B) 1   C) 3   D) 6

    E) 5   F) 9   G) 4   H) 8

    Answer: 6 easy

2. For the curve $y = x^2$, find the slope M_{PQ} of the secant line through the points $P = (1, 1)$ and $Q = (2, 4)$.

    A) 8   B) 2   C) 5   D) 4

    E) 1   F) 9   G) 3   H) 6

    Answer: 3 easy

3. For the curve $y = \sqrt{x} + x$, find the slope M_{PQ} of the secant line through the points
    $P = (1, f(1))$
    and $Q = (4, f(4))$.

    A) 2   B) 1   C) 2/3   D) 3/2

    E) 4/3   F) 3/4   G) 1/2   H) 3

    Answer: 4/3 medium

4. The displacement in meters of a particle moving in a straight line is given by $s = t^2 + t$, where t is measured in seconds. Find the average velocity in meters per second over the time period $[1, 2]$.

    A) 5   B) 3   C) 8   D) 1

    E) 4   F) 6   G) 2   H) 9

    Answer: 4 medium

5. If a ball is thrown into the air with a velocity of 80 ft/s, its height after t seconds is given by $y = 80t - 16t^2$. Find the average velocity in ft/s for the time period beginning when $t = 1$ and lasting 2 seconds.

A) 20   B) 40   C) 30   D) 10

E) 81   F) 36   G) 32   H) 16

Answer: 16 hard

Calculus, 2nd Edition
by James Stewart
Chapter 1, Section 2
The Limit of a Function

1. Find the value of the limit $\lim\limits_{x \to 2} (x^2 + 2)$.

   A) 12   B) 3   C) 2   D) 6

   E) 9   F) 4   G) 8   H) 10

   Answer: 6 easy
   _____

Use the graph below for the following questions:

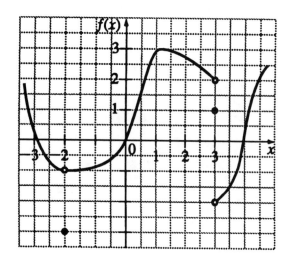

2. For the function whose graph is given above, state limit $\lim\limits_{x \to 1} f(x)$.

   A) −3   B) −2   C) −1   D) 0

   E) 1   F) 2   G) 3   H) Does not exist

   Answer: 3 easy
   _____

3. For the function whose graph is given above, state limit $\lim\limits_{x \to 3^-} f(x)$.

   A) −3   B) −2   C) −1   D) 0

   E) 1   F) 2   G) 3   H) Does not exist

   Answer: 2 easy
   _____

4. For the function whose graph is given above, state limit $\lim\limits_{x \to 3^+} f(x)$.

A) -3 B) -2 C) -1 D) 0

E) 1   F) 2   G) 3   H) Does not exist

Answer: -2 easy
_____

5. For the function whose graph is given above, state limit $\lim\limits_{x \to 3} f(x)$.

A) -3 B) -2 C) -1 D) 0

E) 1   F) 2   G) 3   H) Does not exist

Answer: Does not exist easy
_____

6. For the function whose graph is given above, state limit $f(3)$.

A) -3 B) -2 C) -1 D) 0

E) 1   F) 2   G) 3   H) Does not exist

Answer: 1 easy
_____

7. For the function whose graph is given above, state limit $\lim\limits_{x \to -2^-} f(x)$.

A) -3 B) -2 C) -1 D) 0

E) 1   F) 2   G) 3   H) Does not exist

Answer: -1 easy
_____

8. For the function whose graph is given above, state limit $\lim\limits_{x \to -2^+} f(x)$.

A) -3 B) -2 C) -1 D) 0

E) 1   F) 2   G) 3   H) Does not exist

Answer: -1 easy
_____

9. For the function whose graph is given above, state limit $\lim\limits_{x \to -2} f(x)$.

A) -3 B) -2 C) -1 D) 0

E) 1   F) 2   G) 3   H) Does not exist

Answer: -1 easy
_____

10.   For the function whose graph is given above, state limit $f(-2)$.

A) -3 B) -2 C) -1 D) 0

E) 1   F) 2   G) 3   H) Does not exist

Answer: -3 easy
_____

Calculus, 2nd Edition
by James Stewart
Chapter 1, Section 3
Calculating Limits using the Limit Laws

1. Find the value of the limit $\lim\limits_{x \to 1} \left(x^{17} - x + 3 \right)$.

   A) -4 B) 3 C) -2 D) 1

   E) -8 F) 2 G) 0 H) -16

   Answer: 3 easy

2. Find the value of the limit $\lim\limits_{x \to 2} \dfrac{x^3 - 8}{x - 2}$.

   A) 8 B) 4 C) 12 D) 2

   E) 6 F) 1 G) 0 H) 3

   Answer: 12 medium

3. Find the value of the limit $\lim\limits_{x \to 2^+} \dfrac{|x - 2|}{x - 2}$.

   A) 4 B) -4 C) $1/2$ D) -2

   E) -1 F) 2 G) 1 H) $-1/2$

   Answer: 1 medium

4. Find the value of the limit $\lim\limits_{x \to 2^-} \dfrac{|x - 2|}{x - 2}$.

   A) 1 B) -2 C) $1/2$ D) -4

   E) -1 F) 4 G) $-1/2$ H) 2

   Answer: -1 hard

5. Find the value of the limit $\lim\limits_{x \to 1} \dfrac{x - 1}{\sqrt{x} - 1}$.

   A) $\sqrt{2}$ B) 1 C) -1 D) $-\sqrt{2}$

   E) 2 F) -4 G) 4 H) -2

   Answer: 2 medium

6. Find the value of the limit $\lim\limits_{x \to 1} \dfrac{\sqrt[3]{x} - 1}{\sqrt{x} - 1}$.

A) 1   B) 0   C) 2/3   D) $\sqrt{2/3}$

E) $\sqrt{3/2}$ F) 3/2   G) $\sqrt{3-1}$ H) $\sqrt{2-1}$

Answer: 2/3 hard

7. Let $f(x) = 2x - 1$ if x is rational; $= 1$ if x is irrational. Find the value a for which $\lim\limits_{x \to a} f(x)$ exists.

A) 1/2   B) -2 C) 2   D) $-1/2$

E) 3/4   F) $-3/4$ G) -1 H) 1

Answer: 1 medium

8. Find the value of the limit $\lim\limits_{x \to 1^+} \dfrac{x^2 + x - 2}{|x - x|}$.

A) -3 B) 1   C) 2   D) 0

E) 3   F) 1/2   G) -2 H) does not exist

Answer: 3 hard

9. Find the value of the limit $\lim\limits_{x \to 4^-} \dfrac{|x - 4|}{x - 4}$.

A) ∞ B) 1   C) 0   D) -1

E) 2   F) -3 G) 4   H) does not exist

Answer: -1 medium

10. Find the value of the limit $\lim\limits_{x \to -2} \left[\dfrac{1}{x + 2} + \dfrac{4}{x^2 - 4} \right]$.

A) -4 B) -2 C) $-1/2$ D) $-1/4$

E) 1/4   F) 1/2   G) 2   H) 4

Answer: $-1/4$ hard

11. Find the value of the limit $\lim\limits_{x \to 1} \dfrac{x^3 - 1}{x^2 - 1}$.

A) 1/3   B) 1/6   C) 1/2   D) 3/2

E) 1   F) 1/4   G) 2   H) 3

Answer: 3/2 hard

12. Find the value of the limit $\lim_{x \to 1} \frac{|x| - x}{x - 1}$.

   A) 2   B) 1/2   C) −1/2   D) 1

   E) −1   F) −1/4   G) 1/4   H) 0

   Answer: 0 hard
   _____

13. Find the value of the limit $\lim_{x \to 0^-} x \sin \frac{1}{x}$.

   A) −∞   B) −1/2   C) 1   D) −1

   E) 0   F) sin 1   G) 1/2   H) does not exist

   Answer: 0 medium
   _____

14. Find the value of the limit $\lim_{x \to 3^-} \frac{|x - 3|}{x - 3}$.

   A) −4   B) −2   C) −1   D) 0

   E) 2   F) 4   G) 6   H) does not exist

   Answer: −1 medium
   _____

15. Find the value of the limit $\lim_{h \to 0} \frac{(h - 2)^2 - 4}{h}$.

   A) 2   B) 1   C) −2   D) −4

   E) 1/2   F) 4   G) −1   H) −1/2

   Answer: −4 medium
   _____

16. Find the value of the limit $\lim_{h \to 0} \frac{(h + 2)^2 - 4}{h}$.

   A) 1   B) 2   C) −1   D) −4

   E) 4   F) −1/2   G) 1/2   H) −2

   Answer: 4 medium
   _____

17. Find the value of the limit $\lim_{x \to 3} \frac{1/x - 1/3}{x - 3}$.

   A) 1/12   B) 1/6   C) −1/6   D) 1/3

   E) −1/9   F) −1/12   G) −1/3   H) 1/9

   Answer: −1/9 hard
   _____

18. Find the value of the limit $\lim\limits_{x \to -1} \dfrac{x^2 + 2x + 1}{x^2 - x - 2}$.

   A) 1   B) 2   C) -2 D) 0

   E) -1 F) 1/2   G) $-1/2$ H) does not exist

   Answer: 0 medium
   _____

19. Find the value of the limit $\lim\limits_{x \to 1} \dfrac{x - 1}{x^4 - 1}$.

   A) 0   B) 2   C) 4   D) 8

   E) 1/4   F) 1/8   G) 1/32   H) 1/64

   Answer: 1/4 medium
   _____

20. Find the value of the limit $\lim\limits_{x \to 1} \dfrac{x^4 - 1}{x - 1}$.

   A) 0   B) 1   C) 2   D) 3

   E) 4   F) 5   G)   H) does not exist

   Answer: 4 medium
   _____

21. Find the value of the limit $\lim\limits_{x \to 4} \dfrac{x - 4}{\sqrt{x} - 2}$.

   A) 0   B) 2   C) 4   D) 8

   E) 1/4   F) 1/8   G) 1/32   H) 1/64

   Answer: 4 hard
   _____

22. Find the value of the limit $\lim\limits_{x \to 2} \dfrac{x^2 + x - 6}{x^2 - 5x + 6}$.

   A) -5 B) -3 C) -1 D) 0

   E) 1   F) 3   G) 5   H) 6

   Answer: -5 medium
   _____

23. Find the value of the limit $\lim\limits_{x \to 2} \dfrac{x - 2}{x^4 - 16}$.

   A) 0   B) 2   C) 4   D) 8

   E) 1/4   F) 1/8   G) 1/32   H) 1/64

   Answer: 1/32 medium
   _____

Calculus, 2nd Edition
by James Stewart
Chapter 1, Section 4
Functions and Their Graphs

1. How close to 7 do we have to take x so that $3x + 4$ is within a distance of 0.2 from 25?

    A) 1/21   B) 1/10   C) 1/15   D) 1/14

    E) 1/28   F) 1/7   G) 1/25   H) 1/5

    Answer: 1/15 medium

2. In using the ϵ, δ notation to prove that $\lim_{x \to 2} (2x - 1) = 3$, when ϵ is 1/2, what is the largest value

    that δ can have?

    A) 3/8   B) 7/8   C) 5/8   D) 3/4

    E) 1/8   F) 1   G) 1/4   H) 1/2

    Answer: 1/4 medium

3. In using the ϵ, δ notation to prove that $\lim_{x \to 0} x^2 = 0$, when ϵ is 1/4, what is the largest value that

    δ can have?

    A) 1/2   B) 4   C) 1/4   D) 1

    E) 1/16   F) 2   G) 0   H) 1/8

    Answer: 1/2 medium

4. In using the ϵ, δ notation to prove that $\lim_{x \to 1} x^2 = 1$, when ϵ is 1, what is the largest value that

    δ can have?

    A) 2   B) 1/2   C) $\sqrt{2} - 1$ D) 1/4

    E) $\sqrt{2}/2$ F) 1   G) $\sqrt{2}$ H) $\sqrt{2} + 1$

    Answer: $\sqrt{2} - 1$ hard

5. In using the ϵ, δ notation to prove that $\lim\limits_{x \to 1} \sqrt{x} = 1$, when ϵ is 1, what is the largest value that δ can have?

A) $\sqrt{2}$ B) $1/2$ C) $\sqrt{3}$ D) $\sqrt{3}/3$

E) 1 F) 2 G) $\sqrt{2}/2$ H) 3

Answer: 1 hard

6. For what value of x does the function $(x-1)^2/(x^2-1)$ fail to have a limit?

A) 2 B) $-1/2$ C) 1 D) $\sqrt{2}$

E) $1/2$ F) 0 G) -2 H) -1

Answer: -1 medium

7. Use the ϵ, δ definition of a limit to prove $\lim\limits_{x \to a} c = c$.

Answer: Given $\epsilon > 0$, we need $\delta > 0$ so that if $|x - a| < \delta$ then $|c - c| < \epsilon$. But $|c - c| = 0$,

so this will be true no matter what δ we pick. medium

8. Use the ϵ, δ definition of a limit to prove $\lim\limits_{x \to 4} (5 - 2x) = -3$.

Answer: Given $\epsilon > 0$, we need $\delta > 0$ so that if $|x - 4| < \delta$, then $|(5 - 2x) - (-3)| < \epsilon$

$\Leftrightarrow |-2x + 8| < \epsilon \Leftrightarrow 2|x - 4| < \epsilon \Leftrightarrow |x - 4| < \frac{\epsilon}{2}$. So choose $\delta = \frac{\epsilon}{2}$. Then $|x - 4| < \delta \Rightarrow$

$|(5 - 2x) - (-3)| < \epsilon$. Thus $\lim\limits_{x \to 4} (5 - 2x) = -3$ by the definition of a limit. medium

9. Use the ϵ, δ definition of a limit to prove $\lim\limits_{x \to 0} |x| = 0$.

Answer: Given $\epsilon > 0$, we need $\delta > 0$ so that if $|x - 0| < \delta$ then $||x| - 0| < \epsilon$. But $||x|| = |x|$.

So this is true if we pick $\delta = \epsilon$. medium

10. Use the ϵ, δ definition of a limit to prove $\lim_{x \to 0} x^3 = 0$.

Answer: Given $\epsilon > 0$, we need $\delta > 0$ so that if $|x| < \delta$ then $|x^3 - 0| < \epsilon \Leftrightarrow |x|^3 < \epsilon$

$\Leftrightarrow |x| < \sqrt[3]{\epsilon}$. Take $\delta = \sqrt[3]{\epsilon}$. Then $|x - 0| < \delta \Rightarrow |x^3 - 0| < \delta^3 = \epsilon$. Thus $\lim_{x \to 0} x^3 = 0$

by the definition of a limit. hard

11. Use the ϵ, δ definition of a limit to prove $\lim_{x \to -4} \dfrac{1}{(x+4)^2}$ does not exist.

Answer: Given $\epsilon > 0$, we need $\delta > 0$ so that if $|x - (-2)| < \delta$ then $|(x^2 - 1) - 3| < \epsilon$ or upon

simplifying we need $|x^2 - 4| < \epsilon$ whenever $|x + 2| < \delta$. Notice that if $|x + 2| < 1$, then

$-1 < x + 2 < 1 \Rightarrow -5 < x - 2 < -3 \Rightarrow |x - 2| < 5$. So take $\delta = \min\{\epsilon/5, 1\}$. Then

$|x - 2| < 5$ and $|x + 2| < \epsilon/5$, so $|(x^2 - 1) - 3| = |(x + 2)(x - 2)| =$

$|x + 2||x - 2| < (\epsilon/5)(5) = \epsilon$. Therefore by the definition of a limit $\lim_{x \to -2} x^2 - 1 = 8$.

hard

Calculus, 2nd Edition
by James Stewart
Chapter 1, Section 5
Continuity

1. At what value of x does the function $(x+1)^2/(x^2-1)$ have a removable discontinuity?

    A) −3   B) 3   C) 2   D) −1

    E) 1   F) −4   G) −2   H) 4

    Answer: −1 easy
    _____

2. At what value of x does the function $(x+1)^2/(x^2-1)$ have an infinite discontinuity?

    A) 3   B) −4   C) −2   D) −3

    E) −1   F) 1   G) 2   H) 4

    Answer: 1 medium
    _____

3. At what value of x does the function $|x-2|/(x-2)$ have a jump discontinuity?

    A) −3   B) −1/2   C) 3   D) 1

    E) 2   F) −2   G) 1/2   H) −1

    Answer: 2 medium
    _____

4. At how many values of x is the function $1/(x^3+1)$ discontinuous?

    A) 1   B) 6   C) 0   D) 7

    E) 5   F) 4   G) 2   H) 3

    Answer: 1 medium
    _____

5. Find the distance between the two values of x at which the function $1/(x^2-3x+2)$ is discontinuous.

    A) 3   B) 2   C) 8   D) 1

    E) 5   F) 4   G) 7   H) 6

    Answer: 1 medium
    _____

6. At how many values of x is the function $\sin \frac{1}{x}$ discontinuous?

A) 4   B) 3   C) 2   D) 1

E) 6   F) 5   G) 0   H) infinitely many

Answer: 1 medium

7. At how many values of x is the function $1/(\sin x)$ discontinuous?

A) 5   B) 0   C) 3   D) 4

E) 1   F) 6   G) 2   H) infinitely many

Answer: infinitely many (medium)

8. At what value or values of x is the function

$$f(x) = \begin{cases} |x+1|-1 & \text{if } x < 0 \\ x^2+x & \text{if } 0 \le x < 1 \\ 3-x & \text{if } 1 \le x \end{cases}$$

discontinuous?

A) -1 B) $-1, 0, 1$ C) 1   D) 0

E) $0, 1$ F) $-1, 0$ G) $-1, 1$ H) continuous everywhere

Answer: continuous everywhere (medium)

9. At what value or values of x is the function

$$f(x) = \begin{cases} x+4 & \text{if } x \le -1 \\ x^2 & \text{if } -1 < x < 1 \\ 2-x & \text{if } x \ge 1 \end{cases}$$

discontinuous?

A) -1 B) 0   C) 1   D) $-1, 0$

E) $0, 1$ F) $-1, 1$ G) $-1, 0, 1$ H) continuous everywhere

Answer: -1 medium

10. For all but two choices of a, the function

$$f(x) = \begin{cases} x^3 & \text{if } x \le a \\ x^2 & \text{if } x > a \end{cases}$$

will be discontinuous at the point $x = a$. What values of a will make it continuous?

A) $-2, 2$ B) $1, 2$ C) $-2, 1$ D) $\sqrt{2}, 1$

E) $\sqrt{2}, -1$ F) $-1, 1$ G) $0, 1$ H) $-1, 0$

Answer: $0, 1$ medium

11. At what value or values of x is the function

$$f(x) = \begin{cases} x + 2 & \text{if } x \le -1 \\ x^2 & \text{if } -1 < x < 1 \\ 3 - x & \text{if } x \ge 1 \end{cases}$$

discontinuous?

A) -1 B) 0 C) 1 D) $-1, 0$

E) $0, 1$ F) $-1, 1$ G) $-1, 0, 1$ H) continuous everywhere

Answer: 1 medium

12. For $f(x) = \dfrac{x - 3}{x^2 - 9}$ an infinite discontinuity occurs at what value of x.

A) -1 B) 0 C) 1 D) -3

E) 3 F) $-3, 3$ G) $-3, 1$ H) $3, -1$

Answer: -3 medium

Calculus, 2nd Edition
by James Stewart
Chapter 1, Section 6
Tangents, Velocities, and Other Rates of Change

1. Suppose you drive for 60 miles at 60 miles per hour, then for 60 miles at 30 miles per hour. In miles per hour, what is your average velocity?

    A) 45   B) 40   C) 36   D) 42

    E) 52   F) 55   G) 50   H) 48

    Answer: 40 medium

2. If a ball is thrown into the air with a velocity of 80 ft/s, its height in feet after t seconds is given by $y = 80t - 16t^2$. It will be at maximum height when its instantaneous velocity is zero. Find its average velocity from the time it is thrown ($t = 0$) to the time it reaches its maximum height.

    A) 50   B) 60   C) 100   D) 80

    E) 40   F) 30   G) 48   H) 32

    Answer: 40 hard

3. The curve $f(x) = -x^4 + 2x^2 + x$ has a tangent at the point $(1, 2)$. Is this line tangent to the curve at another point and, if so, where?

    A) $(2, 2)$ B) $(-2, -2)$ C) $(-1, 0)$

    D) $(0, 0)$ E) $(1/2, 15/16)$ F)$(-1/2, -1/16)$

    G) $(1, 2)$ H) tangent at no other point

    Answer: $(-1, 0)$ hard

4. Find an equation of the line tangent to $f(x) = x^2 - 4x$ at the point $(3, -3)$.

    A) $2x - y = 9$ B) $x - 2y = 9$ C) $y - 2x = 9$

    D) $2y - x = 9$ E) $x - y = 9$ F) $y - x = 9$

    G) $2x - y = 4$ H) $x - 2y = 4$

    Answer: $2x - y = 9$ medium

5. Find an equation of the line tangent to the curve $y = x + (1/x)$ at the point $(5, 26/25)$.

A) $24x - 25y = 10$

B) $24x - 25y = -10$

C) $25x - 24y = 10$

D) $25x - 24y = -10$

E) $24x - 24y = 10$

F) $24x - 24y = -10$

G) $25x - 25y = 10$

H) $25x - 25y = -10$

Answer: $24x - 25y = -10$ medium

6. Find the equation of the line(s) with slope $\frac{1}{2}$ tangent to the curve $y = 3x^3$.

A) $y = \frac{1}{2}x - \dfrac{1}{9\sqrt{2}}$

B) $y = \frac{1}{2}x + \dfrac{1}{9\sqrt{2}}$

C) $y = \frac{1}{2}x - \dfrac{1}{9\sqrt{2}}, \quad y = \frac{1}{2}x + \dfrac{1}{9\sqrt{2}}$

D) $y = \frac{1}{2}x - \dfrac{1}{\sqrt{2}}$

E) $y = \frac{1}{2}x + \dfrac{1}{\sqrt{2}}$

F) $y = \frac{1}{2}x - \dfrac{1}{\sqrt{2}}, \quad y = \frac{1}{2}x + \dfrac{1}{\sqrt{2}}$

G) $y = \frac{1}{2}x - \dfrac{1}{9\sqrt{2}}, \quad y = \frac{1}{2}x + \dfrac{1}{\sqrt{2}}$

H) $y = \frac{1}{2}x - \dfrac{1}{\sqrt{2}}, \quad y = \frac{1}{2}x + \dfrac{1}{9\sqrt{2}}$

Answer: $y = \frac{1}{2}x - \dfrac{1}{9\sqrt{2}}, \quad y = \frac{1}{2}x + \dfrac{1}{9\sqrt{2}}$ hard

7. Find the equation of the line which is tangent to the graph of the function $f(x) = 3x^2 - 3x$ and which is parallel to the x–axis.

A) $y = 3/4$

B) $y = -3/4$

C) $y = 1/2$

D) $y = -1/2$

E) $y = 1/4$

F) $y = -1/4$

G) $y = 1/3$

H) $y = -1/3$

Answer: $y = -3/4$ hard

Calculus, 2nd Edition
by James Stewart
Chapter 2, Section 1
Derivatives

1. Let $f(x) = x^2 - 3$. Find $f'(x)$.

   A) $2x$ B) $2x^3 - 3x$ C) $x^3 - 3x$ D) x

   E) $2x - 3$ F) $x - 3$ G) $2x^2 - 3$ H) $x^2 - 4$

   Answer: $2x$ easy

2. Let $f(x) = 3x - 2x^2$. Find $f'(x)$.

   A) $3 - 4x^2$ B) $3x^2 - 4x^3$ C) $3 - 4x$ D) $6 - 4x^2$

   E) $6 - 4x$ F) $3 - x$ G) $3x - x^2$ H) $3 - 2x$

   Answer: $3 - 4x$ easy

3. Find the slope of the tangent line to the curve $y = x^2 + 2x$ at the point $(1, 3)$.

   A) 10   B) 8   C) 6   D) 4

   E) 3   F) 5   G) 1   H) 2

   Answer: 4   easy

4. Find the slope of the tangent line to the curve $y = 1/(1 + x)$ at the point $(1, 1/2)$.

   A) $-1/4$ B) 2   C) -1 D) $1/4$

   E) $-1/2$ F) -2 G) 1   H) $1/2$

   Answer: $-1/4$ medium

5. Find the y-intercept of the tangent line to the curve $y = x^3$ at the point $(2, 8)$.

   A) -4 B) 4   C) -16 D) 12

   E) -8 F) 8   G) -12 H) 16

   Answer: -16 medium

6. Find the y-intercept of the tangent line to the curve $y = \sqrt{x+3}$ at the point $(1, 2)$.

A) 3/4   B) 7/4   C) 1/4   D) 2

E) 1/2   F) 1   G) 3/2   H) 5/4

Answer: 7/4 medium

_____

7. At what value or values of x is the function

$$f(x) = \begin{cases} |x+1| - 1 & \text{if } x < 0 \\ x^2 + x & \text{if } 0 \leq x < 1 \\ 3 - x & \text{if } 1 \leq x \end{cases}$$

not differentiable?

A) -1 B) $-1, 0, 1$ C) 1   D) 0

E) 1, 0   F) $-1, 0$ G) $-1, 1$ H) differentiable everywhere

Answer: $-1, 1$ hard

_____

8. At what value or values of x is the function

$$f(x) = \begin{cases} x + 2 & \text{if } x \leq -1 \\ x^2 & \text{if } -1 < x < 1 \\ 3 - x & \text{if } 1 \leq x \end{cases}$$

not differentiable?

A) -1 B) 0   C) 1   D) $-1, 0$

E) 0, 1   F) $-1, 1$ G) $-1, 0, 1$ H) differentiable everywhere

Answer: $-1, 1$ hard

_____

9. At $x = 0$ the function

$$f(x) = \begin{cases} x^2 & \text{if } x < 0 \\ x & \text{if } x \geq 0 \end{cases}$$

is:

A) differentiable and continuous   B) continuous but not differentiable

C) differentiable but not continuous   D) neither continuous nor differentiable

Answer: continuous but not differentiable (medium)

_____

10.   Where is the greatest integer function $f(x) = [\![x]\!]$ not differentiable?

Answer: $f(x) = [\![x]\!]$ is not continuous at any integer n so f is not differentiable at n by

Theorem 2.8. If a is not an integer, then f is constant on an open interval containing

a, so $f'(a) = 0$. Thus $f'(x) = 0$, x not an integer. medium

11.   A function f is called even if $f(-x) = f(x)$ for all x in its domain and odd if $f(-x) = -f(x)$ for all such x. Prove that the derivative of an even function is an odd function.

Answer: If f is even, then $f'(-x) = \lim_{h \to 0} \dfrac{f(-x+h) - f(-x)}{h} = \lim_{h \to 0} \dfrac{-f(x-h) + f(x)}{h} =$

$-\lim_{h \to 0} \dfrac{f(x-h) - f(x)}{-h}$ [Let $\Delta x = -h$.] $= -\lim_{\Delta x \to 0} \dfrac{f(x+\Delta x) - f(x)}{\Delta x} = -f'(x)$.

Therefore f' is odd. medium

Calculus, 2nd Edition
by James Stewart
Chapter 2, Section 2
Differentiation Formulas

1. Let $f(x) = 1/x^2$. Find $f'(1)$.

    A) $-1/3$ B) -2 C) $1/3$ D) $-1/4$

    E) 1 F) $-1/2$ G) -1 H) $1/4$

    Answer: -2 easy

2. Let $f(x) = \sqrt{3x}$. Find $f'(3)$.

    A) $1/3$ B) $1/2$ C) $-1/3$ D) $-1/4$

    E) -1 F) $1/4$ G) $-1/2$ H) 1

    Answer: $1/2$ medium

3. Let $f(x) = (x+1)^2(x+2)$. Find $f'(0)$.

    A) 6 B) 9 C) 4 D) 12

    E) 8 F) 16 G) 5 H) 20

    Answer: 5 medium

4. Let $f(x) = x^3/(x+2)^2$. Find $f'(-1)$.

    A) 8 B) 5 C) 10 D) 4

    E) 20 F) 9 G) 12 H) 16

    Answer: 5 hard

5. The curve $y = x^3 + x^2 - x$ has two horizontal tangents. Find the difference in height of these two horizontal lines.

    A) $11/9$ B) $22/27$ C) $32/27$ D) $5/3$

    E) $14/9$ F) $4/3$ G) $13/9$ H) $7/3$

    Answer: $32/27$ hard

6. Let $f(x) = {}^5\sqrt{x} + x^{-1.8}$. Find $f'(1)$.

A) 1/5   B) 2/5   C) −8/5   D) −2/5

E) −1/5   F) 4/5   G) 8/5   H) −4/5

Answer: −8/5 medium

7. Passing through the origin $(0, 0)$, there are two lines tangent to the curve $y = x^2 + 1$, one with negative slope, the other with positive slope. Find the value of that positive slope.

A) 1/8   B) 1/2   C) 1/4   D) 1/3

E) 4   F) 1   G) 3   H) 2

Answer: 2 hard

8. If $f(x) = \dfrac{\sqrt{x} - 1}{\sqrt{x} + 1}$, find the value of $f'(4)$.

A) 1/9   B) 1/12   C) 1/15   D) 1/18

E) 1/21   F) 1/24   G) 1/27   H) 1/30

Answer: 1/18 medium

9. At how many different values of x does the curve $y = x^3 + 2x$ have a tangent line parallel to the line $y = x$?

A) 0   B) 1   C) 2   D) 3

E) 4   F) 5   G) 6   H) 7

Answer: 0 medium

10. Find an equation of the tangent line to the curve $y = \frac{1-x}{1+x}$ at $(-2, -3)$.

A) $x + y + 5 = 0$ B) $x + 2y + 8 = 0$ C) $x + 3y + 11 = 0$

D) $2x + y + 7 = 0$ E) $3x + y + 9 = 0$ F) $2x + 3y + 13 = 0$

G) $2x - 3y - 5 = 0$ H) $x - 3y - 7 = 0$

Answer: $2x + y + 7 = 0$ medium

11. There are two lines through the point $(2, -3)$ that are tangent to the parabola $y = x^2 + x$. Find the x-coordinates of the points where these lines touch the parabola.

A) $-2, 0$ B) $1, 3$ C) $-1, 6$ D) $2, 5$

E) $0, 4$ F) $-2, 4$ G) $2, 7$ H) $-1, 5$

Answer: $-1, 5$ hard

12. Given $f(3) = 5$, $f'(3) = 1.1$, $g(3) = -4$ and $g'(3) = 0.7$ find the value of $(f + g)'(3)$.

A) 1.8 B) 0.4 C) -0.9 D) -1.8

E) 0.9 F) -0.4 G) 0.77 H) -0.77

Answer: 1.8 easy

13. Given $f(3) = 5$, $f'(3) = 1.1$, $g(3) = -4$ and $g'(3) = 0.7$ find the value of $(f - g)'(3)$.

A) 1.8 B) 0.4 C) -0.9 D) -1.8

E) 0.9 F) -0.4 G) 0.77 H) -0.77

Answer: 0.4 easy

14. Given $f(3) = 5$, $f'(3) = 1.1$, $g(3) = -4$ and $g'(3) = 0.7$ find the value of $(f \cdot g)'(3)$.

A) 1.8 B) 0.4 C) -0.9 D) -1.8

E) 0.9 F) -0.4 G) 0.77 H) -0.77

Answer: -0.9 easy

15. Given $f(3) = 5$, $f'(3) = 1.1$, $g(3) = -4$ and $g'(3) = 0.7$ find the value of $(f/g)'(3)$.

A) 0.025 B) 0.49375 C) -0.49375 D) -0.025

E) 1.975 F) -1.975 G) 0.5625 H) -0.5625

Answer: -0.49375 easy

Calculus, 2nd Edition
by James Stewart
Chapter 2, Section 3
Rates of Change in the Natural and Social Sciences

1. A stone is thrown into a pond creating a circular wave whose radius increases at the rate of 1 foot per second. In square feet per second, how fast is the area of the circular ripple increasing 3 seconds after the stone hits the water?

   A) π B) 2π C) $\pi/3$ D) 6π

   E) 3π F) $\pi/6$ G) $\pi/12$ H) $\pi/2$

   Answer: 6π easy

2. The mass of a rod varies in such a fashion that the total mass at x meters from the end is x^2 kilograms. Find the density in kg/m at the point 2 meters from the end.

   A) 1   B) 5   C) 3   D) 4

   E) 7   F) 6   G) 2   H) 8

   Answer: 4   easy

3. In a water pipe 2 cm in diameter, the rate of flow at a distance of 0.5 cm from the wall is 6 cm/s. If the flow obeys Poiseuille's law, find the rate of flow at the exact center of the pipe (i.e., 1 cm from the wall).

   A) 6   B) 3   C) 2   D) 10

   E) 8   F) 5   G) 1   H) 12

   Answer: 8   hard

4. A spherical balloon is being inflated in such a fashion that its radius increases at a rate of 1 cm/s. In cm^3/s, how fast is the volume increasing 3 seconds after inflation starts?

   A) 36   B) 12π C) 24   D) 18π

   E) 36π F) 18   G) 12   H) 24π

   Answer: 36π medium

5. A spherical balloon is being inflated in such a fashion that its radius increases at a rate 1 cm/s. In cm^2/s, how fast is the surface area increasing 3 seconds after inflation starts?

A) 18   B) 36π C) 36   D) 12

E) 24π F) 18π G) 24   H) 12π

Answer: 24π medium

_____

6. Suppose the amount of a drug left in the body x hours after administration is $20/(x+1)$ mg. In mg/hr, find the rate of decrease of the drug 4 hours after administration.

A) 3   B) 2   C) 4/5   D) 1/2

E) 5/4   F) 2/5   G) 1   H) 5/2

Answer: 4/5 easy

_____

7. The population of a bacteria colony after t hours is $70 + 5t + 2t^2$. Find the growth rate when $t = 3$.

A) 18   B) 9   C) 7   D) 19

E) 21   F) 13   G) 17   H) 11

Answer: 17 easy

_____

8. A particle moves along a straight line with equation of motion $s = t^2 - 2t$. Find the instantaneous velocity of the particle at time $t = 1$.

A) 1   B) 4   C) 0   D) 3

E) 8   F) 6   G) 5   H) 2

Answer: 0 easy

_____

9. A particle moves along a straight line with equation of motion $s = t^2 - 3t + 2$. Find the value of t at which it reverses its direction.

A) 1/2   B) 0   C) 3/2   D) 2/3

E) 1   F) 2   G) 3/4   H) 4/3

Answer: 3/2 medium

_____

10. A particle moves along a straight line with equation of motion $s = t^3 + 2t$. Find the smallest value of its velocity.

A) 1/2   B) −2   C) 1/2   D) −1/2

E) −3   F) 2   G) 3   H) −1/3

Answer: 2 hard

_____

11. A particle moves according to a law of motion $s = 4t^3 - 9t^2 + 6t + 2$, $t \geq 0$ where s is measured in feet and t in seconds. When is the particle at rest?

A) $t = \frac{1}{2}$ B) $t = 1$ C) $t = 0$ D) $t = \frac{1}{2}, 1$

E) $t = 1, 0$ F) $t = \frac{1}{2}, 0$ G) $t = \frac{1}{2}, 0, 1$ H) $t = \frac{3}{2}$

Answer: $t = \frac{1}{2}, 1$ medium

_____

12. A particle moves according to a law of motion $s = 4t^3 - 9t^2 + 6t + 2$, $t \geq 0$ where s is measured in feet and t in seconds. When is the particle moving in a positive direction?

A) $0 \leq t < \frac{1}{2}$ B) $t > 1$ C) $t \geq \frac{3}{2}$

D) $0 \leq t < \frac{1}{2}$ or $t > 1$ E) $0 \leq t < \frac{1}{2}$ or $t > \frac{3}{2}$ F) $1 < t < \frac{3}{2}$

G) $0 \leq t < \frac{1}{2}$ or $1 < t < \frac{3}{2}$ H) $0 \leq t < \frac{3}{2}$

Answer: $0 \leq t < \frac{1}{2}$ or $t > 1$ medium

_____

13. A particle moves according to a law of motion $s = \sqrt{t}(5 - 5t + 2t^2)$, $t \geq 0$ where s is measured in feet and t in seconds. When is the particle moving in a positive direction?

A) $0 \leq t < \frac{1}{2}$ B) $t > 1$ C) $t \geq \frac{3}{2}$

D) $0 \leq t < \frac{1}{2}$ or $t > 1$ E) $0 \leq t < \frac{1}{2}$ or $t > \frac{3}{2}$ F) $1 < t < \frac{3}{2}$

G) $0 \leq t < \frac{1}{2}$ or $1 < t < \frac{3}{2}$ H) $0 \leq t < \frac{3}{2}$

Answer: $0 \leq t < \frac{1}{2}$ or $t > 1$ medium

_____

14. A particle moves according to a law of motion $s = \sqrt{t}\,(5 - 5t + 2t^2)$, $t \geq 0$ where s is measured in feet and t in seconds. When is the particle at rest?

A) $t = \frac{1}{2}$

B) $t = 1$

C) $t = 0$

D) $t = \frac{1}{2}, 1$

E) $t = 1, 0$

F) $t = \frac{1}{2}, 0$

G) $t = \frac{1}{2}, 0, 1$

H) $t = \frac{3}{2}$

Answer: $t = \frac{1}{2}, 1$ medium

Calculus, 2nd Edition
by James Stewart
Chapter 2, Section 4
Derivatives of Trigonometric Functions

1. Find the value of the limit $\lim\limits_{x \to 0} \frac{\cos x - 1}{2x}$.

    A) 3   B) 0   C) 1/3   D) 4

    E) 1/4   F) 2   G) 1/2   H) 1

    Answer: 0 easy
    _____

2. Find the value of the limit $\lim\limits_{x \to 0} \frac{\sin x}{2x}$.

    A) 1/3   B) 3   C) 0   D) 1

    E) 1/2   F) 1/4   G) 4   H) 2

    Answer: 1/2 easy
    _____

3. Find the value of the limit $\lim\limits_{x \to 0} \frac{\tan 2x}{x}$.

    A) 4   B) 1   C) 1/4   D) 2

    E) 1/2   F) 0   G) 1/3   H) 3

    Answer: 2 medium
    _____

4. Let $f(x) = x^2 \cos x$. Find $f'(0)$.

    A) 1/2   B) 2   C) 0   D) 1

    E) 1/3   F) 4   G) 1/4   H) 3

    Answer: 0 medium
    _____

5. Let $f(x) = x \tan x$. Find $f'(\pi/4)$.

    A) $1 + \pi/4$ B) $\pi/4$ C) $\pi/4 - 1$ D) $1 - \pi/4$

    E) $\pi/2 - 1$ F) $1 - \pi/2$ G) $1 + \pi/2$ H) $\pi/2$

    Answer: $1 + \pi/2$ medium
    _____

6. Let $f(x) = \sin^2 x$. Find $f'(\pi/4)$.

A) $\sqrt{2}$ B) $\sqrt{2}/3$ C) 2   D) 1

E) $1/2$ F) 0   G) $\sqrt{3}/4$ H) $\sqrt{2}/2$

Answer: 1 medium

_____

7. Let $f(x) = \dfrac{x}{\sin x}$. Find $f'(\pi/4)$.

A) $1 + \pi/4$ B) $1 + \pi/2$ C) $\sqrt{2}(1 + \pi/2)$

D) $1 - \pi/2$ E) $1 - \pi/4$ F) $\sqrt{2}(1 + \pi/4)$

G) $\sqrt{2}(1 - \pi/2)$ H) $\sqrt{2}(1 - \pi/4)$

Answer: $\sqrt{2}(1 - \pi/4)$ hard

_____

8. Find the value of the limit $\displaystyle\lim_{x \to 0} \dfrac{x^2}{\tan^2 \pi x}$.

A) π B) π^2 C) 0   D) 1

E) $1/\pi^2$ F) $1/\pi$ G) ∞ H) does not exist

Answer: $1/\pi^2$ hard

_____

9. Find the value of the limit $\displaystyle\lim_{x \to 0^+} \dfrac{\sin 2x \cos 4x}{\tan 3x}$.

A) 0   B) $2/3$ C) $4/3$ D) 2

E) $8/3$ F) $10/3$ G) 4   H) ∞

Answer: $2/3$ hard

_____

10. Find the value of the limit $\displaystyle\lim_{t \to 0} \dfrac{\sin 6t}{\sin 4t}$.

A) 0   B) 0.5   C) 1   D) 1.5

E) 2   F) 2.5   G) 3   H) does not exist

Answer: 1.5 medium

_____

11. Find the value of the limit $\displaystyle\lim_{x \to 0} \dfrac{\sin 3x}{\tan 6x}$.

A) 0   B) 0.5   C) 1.0   D) 1.5

E) 2.0   F) 2.5   G) 3.0   H) does not exist

Answer: 0.5 medium

_____

12. Prove that $\frac{d}{dx}(\sec x) = \sec x \tan x$.

Answer: $\frac{d}{dx}(\sec x) = \frac{d}{dx}\left(\frac{1}{\cos x}\right) = \frac{(\cos x)(0) - 1(-\sin x)}{\cos^2 x} = \frac{\sin x}{\cos^2 x} = \frac{1}{\cos x} \cdot \frac{\sin x}{\cos x} = \sec x \tan x$.

medium

13. Prove, using the definition of a derivative, that if $f(x) = \cos x$, then $f'(x) = -\sin x$.

Answer: $f(x) = \cos x \Rightarrow f'(x) = \lim_{h \to 0} \frac{f(x+h) - f(x)}{h} = \lim_{h \to 0} \frac{\cos(x+h) - \cos x}{h} =$

$\lim_{h \to 0} \frac{\cos x \cos h - \sin x \sin h - \cos x}{h} = \lim_{h \to 0} \left(\cos x \frac{\cos h - 1}{h} - \sin x \frac{\sin h}{h}\right) =$

$\cos x \lim_{h \to 0} \frac{\cos h - 1}{h} - \sin x \lim_{h \to 0} \frac{\sin h}{h} = (\cos x)(0) - (\sin x)(1) = -\sin x$. medium

Calculus, 2nd Edition
by James Stewart
Chapter 2, Section 5
The Chain Rule

1. Let $f(x) = (x+1)^4$. Find $f'(1)$.

   A) 4   B) 12   C) 8   D) 32

   E) 6   F) 16   G) 24   H) 2

   Answer: 32 easy

2. Let $f(x) = (x^2+1)^4$. Find $f'(0)$.

   A) 0   B) 28   C) 4   D) 32

   E) 16   F) 24   G) 12   H) 8

   Answer: 0 easy

3. Let $f(x) = (x+1)^2(x+2)^3$. Find $f'(0)$.

   A) 4   B) 24   C) 28   D) 12

   E) 16   F) 8   G) 32   H) 6

   Answer: 28 medium

4. Let $f(x) = \sin^2(2x)$. Find $f'(\pi/6)$.

   A) $\sqrt{2}$ B) $\sqrt{3}/4$ C) 0   D) $\sqrt{2}/2$

   E) $\sqrt{2}/4$ F) $\sqrt{3}$ G) $\sqrt{3}/2$ H) 1

   Answer: $\sqrt{3}$ medium

5. Let $f(x) = \sqrt{\sin x + x^3 + 1}$. Find $f'(0)$.

   A) 3   B) 0   C) 1   D) 4

   E) 1/4   F) 2   G) 1/3   H) 1/2

   Answer: 1/2 medium

6. Let $f(x) = \sqrt{x + \sqrt{x}}$. Find $f'(1)$.

   A) $3\sqrt{2}/8$ B) $\sqrt{2}/4$ C) $1/4$ D) $1/8$

   E) $\sqrt{2}/2$ F) $1/2$ G) $\sqrt{2}$ H) 1

   Answer: $3\sqrt{2}/8$ hard
   _____

7. Find the slope of the line tangent to the curve $y = (x^2 + 1)^3$ at the point $(1, 8)$.

   A) 32   B) 4   C) 24   D) 16

   E) 28   F) 6   G) 12   H) 8

   Answer: 24   medium
   _____

8. If $f(x) = \sin\left[\dfrac{\pi}{\sqrt{x^2 + 5}}\right]$, find the value of $f'(2)$.

   A) $\pi/9$ B) $\pi/5$ C) 0   D) $-\pi/27$

   E) $1/2$ F) $\pi\sqrt{3}/2$ G) $-\pi/9$ H) $-\sqrt{3}/2$

   Answer: $-\pi/27$ medium
   _____

9. If $f(x) = 4\sqrt{x - \sqrt{x}}$, find the value of $f'(4)$.

   A) $\dfrac{1}{2\sqrt{2}}$ B) $\dfrac{1}{4\sqrt{2}}$ C) $\dfrac{3}{4\sqrt{2}}$ D) $\dfrac{3}{2\sqrt{2}}$

   E) $\dfrac{3}{\sqrt{2}}$ F) $\dfrac{5}{\sqrt{2}}$ G) $\dfrac{5}{2\sqrt{2}}$ H) $\dfrac{5}{4\sqrt{2}}$

   Answer: $3/(2\sqrt{2})$ hard
   _____

10. Find an equation for the tangent line to the curve $y = \dfrac{8}{\sqrt{4 + 3x}}$ at the point $(4, 2)$.

    A) $6x + y = 26$ B) $4x + 2y = 20$ C) $3x - 4y = 4$

    D) $7x + 18y = 64$ E) $5x + 21y = 62$ F) $4x + 15y = 46$

    G) $3x + 16y = 44$ H) $2x - y = 6$

    Answer: $3x + 16y = 44$ medium
    _____

11. If $f(x) = \sqrt{2x + \sqrt{x}}$, find the value of $f'(1)$.

    A) $\sqrt{2}/4$ B) $\sqrt{2}/8$ C) $3\sqrt{2}/8$ D) $3\sqrt{2}/4$

    E) $3\sqrt{2}/2$ F) $5/(4\sqrt{3})$ G) $4\sqrt{2}$ H) $\sqrt{2}/4$

    Answer: $5/(4\sqrt{3})$ hard
    _____

12. If $f(x) = \sqrt{x + \sqrt{x + \sqrt{x}}}$ find the value of $f'(1)$.

A) $\sqrt{1 + \sqrt{2}}$

B) $\dfrac{1}{\sqrt{1 + \sqrt{2}}}$

C) $\dfrac{1}{2\sqrt{1 + \sqrt{2}}}$

D) $\dfrac{1}{\sqrt{2} + \sqrt{1 + \sqrt{2}}}$

E) $\dfrac{3}{8\sqrt{2}\sqrt{1 + \sqrt{2}}}$

F) $\dfrac{1 + \dfrac{1}{\sqrt{2}}}{2\sqrt{1 + \sqrt{2}}}$

G) $\dfrac{1 + \dfrac{3}{\sqrt{2}}}{2\sqrt{1 + \sqrt{2}}}$

H) $\dfrac{1 + \dfrac{3}{4\sqrt{2}}}{2\sqrt{1 + \sqrt{2}}}$

Answer: $\dfrac{1 + \dfrac{3}{4\sqrt{2}}}{2\sqrt{1 + \sqrt{2}}}$ hard

13. Let $F(x) = \sin(g(x))$, where g is differentiable. Find $F'(x)$.

A) $g(x)\cos x$

B) $-g(x)\sin x$

C) $-g'(x)\sin x$

D) $g'(x)\cos x$

E) $g'(x)\cos(g(x))$

F) $g'(x)\sin(g(x))$

G) $\cos(g(x))$

H) $\cos(g'(x))$

Answer: $g'(x)\cos(g(x))$ medium

14. Suppose that $h(x) = f(g(x))$ and $g(3) = 6$, $g'(3) = 4$, $f'(3) = 2$, $f'(6) = 7$. Find the value of $h'(3)$.

A) 4

B) 8

C) 12

D) 16

E) 20

F) 24

G) 28

H) 32

Answer: 28 medium

15. Suppose that $F(x) = f(g(x))$ and $g(3) = 5$, $g'(3) = 3$, $f'(3) = 1$, $f'(5) = 4$. Find the value of $F'(3)$.

A) 3

B) 4

C) 7

D) 9

E) 12

F) 15

G) 17

H) 20

Answer: 12 medium

16. Suppose that $w = u \circ v$ and $u(0) = 1$, $v(0) = 2$, $u'(0) = 3$, $u'(2) = 4$, $v'(0) = 5$ and $v'(2) = 6$.

Find $w'(0)$.

A) 5   B) 10   C) 15   D) 20

E) 25   F) 30   G) 35   H) 40

Answer: 20   medium

Calculus, 2nd Edition
by James Stewart
Chapter 2, Section 6
Implicit Differentiation

1. If $x^2 + y^2 = 25$, find the value of dy/dx at the point $(3, 4)$.

   A) 3/5   B) −3/5   C) −4/5   D) 3/4

   E) 0   F) 1   G) 4/5   H) −3/4

   Answer: −3/4 easy

2. If $\sqrt{x+y} + \sqrt{x-y} = 4$, find the value of dy/dx at the point $(5, 4)$.

   A) 4   B) 1/4   C) 2   D) −4

   E) −2   F) 1/2   G) −1/4   H) −1/2

   Answer: 2 hard

3. If $x^2 + xy + y^2 = 7$, find the value of dy/dx at the point $(1, 2)$.

   A) −3/5   B) −3/4   C) 3/5   D) 4/5

   E) −4/5   F) 3/4   G) 1   H) 0

   Answer: −4/5 medium

4. If $\sqrt{x} + \sqrt{y} = 3$, find the value of dx/dy at the point $(4, 1)$.

   A) −3   B) 0   C) 2   D) −1

   E) 1   F) 3   G) −2   H) 4

   Answer: −2 medium

5. If $\sin y = x$, find the value of dy/dx at the point $(1/2, \pi/6)$.

   A) $2\sqrt{3}/3$ B) −2   C) −1/2   D) $\sqrt{2}$

   E) 2   F) 1/2   G) $-2\sqrt{3}/3$ H) $-\sqrt{2}$

   Answer: $2\sqrt{3}/3$ hard

6. Find the y-intercept of the tangent to the ellipse $x^2 + 3y^2 = 1$ at the point $(1/2, 1/2)$.

A) 3   B) $-1/\sqrt{3}$ C) 1/3   D) $-1/3$

E) -1 F) -3 G) $1/\sqrt{3}$ H) 2/3

Answer: 2/3   medium

7. Find the slope of the tangent to the curve $xy^2 + x^2y = 2$ at the point $(1, 1)$.

A) 5   B) -1 C) -3 D) 1

E) -5 F) 0   G) 3   H) 4

Answer: -1 medium

8. Find the slope of the tangent line to the curve $2x^3 + 2y^3 - 9xy = 0$ at the point $(1, 2)$.

A) 3   B) 9   C) 9/2   D) 1/3

E) 4/5   F) 18/5   G) 7   H) $-18/5$

Answer: 4/5   medium

9. If $x^2 - xy + y^3 = 8$, find an expression for dy/dx.

A) $\dfrac{y - 2x}{3y^2 - x}$ B) $-\dfrac{2xy}{x + 3y^2}$ C) $-\dfrac{3x^2 + y}{2x - y}$ D) $\dfrac{x^2 - y^2}{x^2 + y^2}$

E) $-\dfrac{3xy^2}{3y^2 + 2x}$ F) $\dfrac{x + y}{x^2 + 2y}$ G) $-\dfrac{2x^2y}{x + 6y^2}$ H) $\dfrac{xy}{x^2 - y^2}$

Answer: $\dfrac{y - 2x}{3y^2 - x}$ medium

10. Let $y = f(x)$. If $xy^3 + xy = 6$ and $f(3) = 1$, find $f'(3)$.

A) 0   B) 1   C) 2   D) 1/3

E) -4 F) 1/5   G) $-1/6$ H) 8

Answer: $-1/6$ medium

11. If $x \cos y = y \cos x$, find the value of dy/dx when $x = 0$ and $y = 0$.

A) -2 B) -1 C) 0   D) 1

E) 2   F) 1/3   G) $-1/2$ H) 1/4

Answer: 1   medium

12. Find the slope of the tangent line to the curve $x^3 + y^3 = 6xy$ at $(3, 3)$.

A) -3 B) -2 C) -1 D) 0

E) 1 F) 2 G) 3 H) 4

Answer: -1 medium

13. If $x \cos y + y \cos x = 1$, find an expression for dy/dx.

A) $\dfrac{y \sin x - \cos y}{\cos x - x \sin y}$ B) $\dfrac{\cos x + \sin y}{\cos x - \sin y}$ C) $\dfrac{\cos x + x \sin y}{\sin x + y \sin y}$

D) $\dfrac{x \sin x + \sin y}{x \cos x - \cos y}$ E) $\dfrac{y - \cos x \sin y}{\sin x + \cos y}$ F) $\dfrac{x \sin x - \cos y}{y \sin x + \cos y}$

G) $\dfrac{\cos x - \sin y}{\cos x + x \sin y}$ H) $\dfrac{\sin x - \cos y}{y \cos x + x \sin y}$

Answer: $\dfrac{y \sin x - \cos y}{\cos x - x \sin y}$ medium

14. Find the derivative $\dfrac{dp}{dt}$ for $p = \dfrac{mv}{\sqrt{1 - (v^2/c^2)}}$, where m and c are constants.

A) $\dfrac{dp}{dt} = \dfrac{m}{\left(1 - (v^2/c^2)\right)^{3/2}} \dfrac{dv}{dt}$ B) $\dfrac{dp}{dt} = \dfrac{m}{\left(1 + (v^2/c^2)\right)^{3/2}} \dfrac{dv}{dt}$

C) $\dfrac{dp}{dt} = \dfrac{m}{\left(1 - (c^2/v^2)\right)^{3/2}} \dfrac{dv}{dt}$ D) $\dfrac{dp}{dt} = \dfrac{m}{\left(1 + (c^2/v^2)\right)^{3/2}} \dfrac{dv}{dt}$

E) $\dfrac{dp}{dt} = \dfrac{m}{\left(1 - (v^2/c^2)\right)^{5/2}} \dfrac{dv}{dt}$ F) $\dfrac{dp}{dt} = \dfrac{mv}{\left(1 - (v^2/c^2)\right)^{3/2}} \dfrac{dv}{dt}$

G) $\dfrac{dp}{dt} = \dfrac{mv}{\left(1 - (v^2/c^2)\right)^{5/2}} \dfrac{dv}{dt}$ H) $\dfrac{dp}{dt} = \dfrac{v}{\left(1 - (v^2/c^2)\right)^{3/2}} \dfrac{dv}{dt}$

Answer: $\dfrac{dp}{dt} = \dfrac{m}{\left(1 - (v^2/c^2)\right)^{3/2}} \dfrac{dv}{dt}$ hard

15. Find the equation of the line normal to the curve defined by the equation $x^3y^4 - 5 = x^3 - x^2 + y$ at the point $(2, -1)$.

A) $33x + 4y - 62 = 0$

B) $33x - 4y - 62 = 0$

C) $33x + 4y + 62 = 0$

D) $33x - 4y + 62 = 0$

E) $4x - 33y - 41 = 0$

F) $4x + 33y - 41 = 0$

G) $4x - 33y + 41 = 0$

H) $4x + 33y + 41 = 0$

Answer: $33x + 4y - 62 = 0$ medium

16. What is the slope of the tangent line to the curve $xy^3 + y - 1 = 0$ at the point $(0, 1)$?

A) -3

B) -2

C) -1

D) 0

E) 1

F) 2

G) 3

H) 4

Answer: -1 medium

17. Find the value of the derivative of $x^2 + xy^2 = 6$ when $x = 2$ and $y > 0$.

A) $-5/4$

B) $-4/5$

C) $-3/4$

D) 0

E) $5/4$

F) $4/5$

G) $3/4$

H) $1/2$

Answer: $-5/4$ medium

18. Find $\dfrac{dy}{dx}$ if $\sin(2x + 3y) = 3xy + 5y - 2$.

A) $\dfrac{dy}{dx} = \dfrac{3y - 2\cos(2x + 3y)}{3\cos(2x + 3y) - 3x - 5}$

B) $\dfrac{dy}{dx} = \dfrac{3y + 2\cos(2x + 3y)}{3\cos(2x + 3y) - 3x - 5}$

C) $\dfrac{dy}{dx} = \dfrac{3y - 2\cos(2x + 3y)}{3\cos(2x + 3y) + 3x - 5}$

D) $\dfrac{dy}{dx} = \dfrac{3y - 2\cos(2x + 3y)}{3\cos(2x + 3y) - 3x + 5}$

E) $\dfrac{dy}{dx} = \dfrac{3y + 2\cos(2x + 3y)}{3\cos(2x + 3y) + 3x - 5}$

F) $\dfrac{dy}{dx} = \dfrac{3y - 2\cos(2x + 3y)}{3\cos(2x + 3y) + 3x + 5}/5$

G) $\dfrac{dy}{dx} = \dfrac{3y + 2\cos(2x + 3y)}{3\cos(2x + 3y) - 3x + 5}$

H) $\dfrac{dy}{dx} = \dfrac{3y + 2\cos(2x + 3y)}{3\cos(2x + 3y) + 3x + 5}$

Answer: $\dfrac{dy}{dx} = \dfrac{3y - 2\cos(2x + 3y)}{3\cos(2x + 3y) - 3x - 5}$ medium

1.   Let $f(x) = \dfrac{1}{x^2}$. Find $f''(1)$.

A) 1   B) 4   C) 2   D) 6

E) 3   F) 0   G) 5   H) 8

Answer: 6 medium

2.   Let $f(x) = x^3$. Find $f''(1)$.

A) 8   B) 4   C) 2   D) 3

E) 0   F) 6   G) 5   H) 1

Answer: 6 easy

3.   Let $f(x) = x \sin x$. Find $f''(0)$.

A) $\sqrt{2}$ B) $\sqrt{2}/2$ C) 2   D) $-\sqrt{3}$

E) $-\sqrt{2}$ F) -2 G) $\sqrt{3}$ H) $-\sqrt{2}/2$

Answer: 2 medium

4.   Let $y = \dfrac{x}{(x+1)}$. Find $f''(0)$.

A) -2 B) 4   C) 8   D) -6

E) -8 F) -4 G) 6   H) 2

Answer: -2 medium

5.   Let $f(x) = x^5$. Find $f^{(5)}(3)$.

A) 90   B) 360   C) 30   D) 60

E) 20   F) 72   G) 120   H) 240

Answer: 120 easy

6. Let $x^2 + y^2 = 25$. Find the value of d^2y/dx^2 at the point $(3, 4)$.

A) 7/16   B) $-7/16$ C) 7/64   D) $-7/64$

E) $-25/16$ F) 25/16   G) $-25/64$ H) 25/64

Answer: $-25/64$ hard

7. A particle moves along a straight line with equation of motion $s = t^3 + t^2$. Find the value of t at which the acceleration is equal to zero.

A) $-2/3$ B) $-1/3$ C) 2/3   D) 1/3

E) $-1/2$ F) 1/2   G) $-3/2$ H) 3/2

Answer: $-1/3$ medium

8. If $f(x) = g(g(x))$, find an expression for $f''(x)$ in terms of $g(x)$ and its derivatives.

A) $2g''(x) \cdot g(x) + 2(g'(x))^2$ B) $g''(g'(x))$

C) $g''(g(x)) \cdot (g'(x))^2 + g'(g(x)) \cdot g''(x)$ D) $g''(g'(x)) \cdot g'(x)$

E) $2g'(x) \cdot g'(x) + 2g''(g(x))$ F) $g'(g'(x))g''(x) + (g'(x))^2$

G) $g'(g''(x)) \cdot g'(x) + (g''(x))^2$ H) $g'(g'(x))g''(x) + (g'(x))^2$

Answer: $g''(g(x)) \cdot (g'(x))^2 + g'(g(x)) \cdot g''(x)$ hard

9. If $x^2 + y^2 = 1$, find an expression for y''.

A) $-2x/y^5$ B) x/y^3 C) $-3x/y^4$ D) $(x^3 - y^3)/y^6$

E) $(x^3 + y^3)/y^6$ F) $-1/y^3$ G) $6x^2/y^5$ H) $-3x^3/y^4$

Answer: $-1/y^3$ hard

10. If $f(x) = \cos 2x$, find the value of $f^{(8)}(0)$.

A) 1   B) -1 C) 2   D) 0

E) 256   F) -256 G) 128   H) -128

Answer: 256 medium

11. If $g(x) = \cos 2x$, find the value of $g^{(6)}(0)$.

A) 0   B) 2   C) -2 D) 8

E) -8 F) 64   G) -64 H) 128

Answer: -64 medium

12. If $f(x) = \tan x$, find an expression for $f'''(x)$.

A) $\sec^2 x \tan^2 x$

B) $2 \sec^4 x$

C) $2 \tan^4 x$

D) $4 \sec^4 x \tan^2 x$

E) $2 \sec^2 x \tan^2 x + 4 \tan^4 x$

F) $4 \sec^2 x \tan^2 x + \tan^4 x$

G) $2 \sec^2 x \tan^2 x + 4 \sec^4 x$

H) $4 \sec^2 x \tan^2 x + 2 \sec^4 x$

Answer: $4 \sec^2 x \tan^2 x + 2 \sec^4 x$ hard

13. If $x^4 + y^4 = 1$, find an expression for $\dfrac{d^2 y}{dx^2}$.

A) y^2/x^5

B) $-x^2/y^2$

C) $-2x^4/y^4$

D) $-2x^2/y^5$

E) $-3x^2/y^7$

F) $-5x^4/y^9$

G) $-6x^4/y^5$

H) $6x^4/y^7$

Answer: $-3x^2/y^7$ hard

14. If $x^3 + y^3 = 1$, find an expression for y''.

A) $-2x/y^5$

B) x/y^3

C) $-3x/y^4$

D) $(x^3 - y^3)/y^6$

E) $(x^3 + y^3)/y^6$

F) $-x^2/y^3$

G) $6x^2/y^5$

H) $-3x^3/y^4$

Answer: $-2x/y^5$ hard

15. Use implicit differentiation to find y'' if $2xy = y^2$.

A) $\dfrac{y^2 - 2xy}{(y-x)^3}$

B) $\dfrac{y^2 + 2xy}{(y-x)^3}$

C) $\dfrac{y^2 - 2xy}{(y+x)^3}$

D) $\dfrac{y^2 + 2xy}{(y+x)^3}$

E) $\dfrac{y^2 - xy}{(y-x)^3}$

F) $\dfrac{y^2 + xy}{(y-x)^3}$

G) $\dfrac{y^2 - xy}{(y+x)^3}$

H) $\dfrac{y^2 + xy}{(y+x)^3}$

Answer: $\dfrac{y^2 - 2xy}{(y-x)^3}$ medium

16. Let $f(x) = x/(x-1)$ find a formula for $f^{(n)}(x)$.

A) $f^{(n)} = (-1)^n n!(x-1)^{-(n+1)}$

B) $f^{(n)} = n!(x-1)^{-(n+1)}$

C) $f^{(n)} = (-1)^n n!(x-1)^{-n}$

D) $f^{(n)} = (-1)^n n!(x-1)^{-(n-1)}$

E) $f^{(n)} = (-1)^{n+1} n!(x-1)^{-(n+1)}$

F) $f^{(n)} = (-1)^n n!(x-1)^{n+1}$

G) $f^{(n)} = n!(x-1)^{-n+1}$

H) $f^{(n)} = (-1)^n n!(x+1)^{-(n+1)}$

Answer: $f^{(n)} = (-1)^n n!(x-1)^{-(n+1)}$ hard

Calculus, 2nd Edition
by James Stewart
Chapter 2, Section 8
Related Rates

1. The length of a rectangle is increasing at the rate of 2 feet per second, while the width is increasing at the rate of 1 foot per second. When the length is 5 feet and the width is 3 feet, how fast, in square feet per second, is the area increasing?

    A) 5   B) 11   C) 6   D) 10

    E) 15   F) 20   G) 8   H) 12

    Answer: 11 medium

2. The length of a rectangle is decreasing at the rate of 1 foot per second, but the area remains constant. If the length is 10 feet and the width is 5 feet, in feet per second, how fast is its width increasing?

    A) 1/2   B) 4   C) 1/10   D) 2

    E) 5   F) 1/4   G) 10   H) 1/5

    Answer: 1/2 hard

3. A ladder 10 feet long is leaning against a wall, with the foot of the ladder 8 feet away from the wall. If the foot of the ladder is being pulled away from the wall at 3 feet per second, how fast in feet per second is the top of the ladder sliding down the wall?

    A) 6   B) 1   C) 3   D) 10

    E) 5   F) 8   G) 2   H) 4

    Answer: 4 medium

4. A cube is 4 feet on each edge, with each edge increasing 1 foot per second. In cubic feet per second, what is the rate of increase in volume?

    A) 56   B) 40   C) 42   D) 72

    E) 64   F) 48   G) 36   H) 96

    Answer: 48 easy

5. Two cars are each 100 miles away from the town of Tucumcari, one directly to the north and the other directly to the east. The car to the north is heading toward the town at 60 miles per hour, while the one to the east is heading toward the town at 30 miles per hour. In miles per hour, how fast are the cars approaching each other?

    A) $200\sqrt{2}$ B) 100   C) $100\sqrt{2}$ D) 200

    E) $50\sqrt{2}$ F) $45\sqrt{2}$ G) 45   H) 50

    Answer: $45\sqrt{2}$ hard

6. A kite is flying 100 feet above the ground at the end of a string 125 feet long. The girl flying the kite lets out the string at a rate of 1 foot per second. If the kite remains 100 feet above the ground, how many feet per second is its horizontal distance from the girl increasing?

    A) 5/3   B) 3   C) 4/3   D) 3/5

    E) 5   F) 5/4   G) 4/5   H) 4

    Answer: 5/3 hard

7. A student 5 feet tall is 10 feet away from a lamppost 15 feet tall. She is walking away from the lamppost at 2 feet per second. How fast is the tip of her shadow moving away from the foot of the lamppost?

    A) 5/2   B) 3/2   C) 3   D) 2/3

    E) 1/3   F) 4   G) 6   H) 1/2

    Answer: 3 hard

8. A plane flying horizontally at an altitude of 1 km and a speed of 500 km/h passes directly over a radar station. Find the rate (in km/h) at which the distance from the plane to the station is increasing when it is 2 km away from the station.

    A) $125\sqrt{3}$ B) $175\sqrt{3}$ C) $250\sqrt{3}$ D) $275\sqrt{3}$

    E) $125\sqrt{2}$ F) $175\sqrt{2}$ G) $250\sqrt{2}$ H) $275\sqrt{2}$

    Answer: $250\sqrt{3}$ medium

9. A lighthouse is on a small island 3 km away from the nearest point P on a straight shoreline, and its light makes 4 revolutions per minute. At what rate (in km/min) is the beam of light moving along the shoreline when it is 1 km away from P?

A) 20π B) $70\pi/3$ C) $80\pi/3$ D) $100\pi/3$

E) $110\pi/3$ F) 40π G) $130\pi/3$ H) $140\pi/3$

Answer: $80\pi/3$ hard

---

10. A cylindrical can is undergoing a transformation in which the radius and height are varying continuously with time t. The radius is increasing at 4 in/min, while the height is decreasing at 10 in/min. Is the volume increasing or decreasing, and at what rate, when the radius is 3 inches and the height is 5 inches?

A) increasing, 30π cubic inches per minute   B) increasing, 20π cubic inches per minute

C) increasing, 10π cubic inches per minute   D) increasing, 40π cubic inches per minute

E) decreasing, 30π cubic inches per minute   F) decreasing, 20π cubic inches per minute

G) decreasing, 10π cubic inches per minute   H) decreasing, 40π cubic inches per minute

Answer: increasing, 30π cubic inches per minute (medium)

---

11. A frugal young man has decided to extract one of his teeth by tying a stout rubber band from his tooth to the chain on a garage door opener which runs on a horizontal track 3 feet above his mouth. If the garage door opener moves the chain at 1/4 ft/s, how fast is the rubber band expanding when it is stretched to a length of 5 feet?

A) 1/5 ft/s   B) 1/2 ft/s   C) 1/3 ft/s   D) 1/4 ft/s

E) 1/6 ft/s   F) 2/5 ft/s   G) 2/3 ft/s   H) 3/5 ft/s

Answer: 1/5 ft/s medium

---

12. Two straight roads intersect at right angles in Newtonville. Car A is on one road moving toward the intersection at a speed of 50 miles/h. Car B is on the other road moving away from the intersection at a speed of 30 miles/h. When car A is 2 miles from the intersection and car B is 4 miles from the intersection how fast is the distance between the cars changing.

A) $\sqrt{20}$ miles/h.   B) $\sqrt{10}$ miles/h.   C) $\sqrt{15}$ miles/h.

D) $\sqrt{30}$ miles/h.   E) $2\sqrt{20}$ miles/h.   F) $2\sqrt{10}$ miles/h.

G) $2\sqrt{15}$ miles/h.   H) $2\sqrt{30}$ miles/h.

Answer: $\sqrt{20}$ miles/h. medium

---

13. The length of a rectangle is increasing at the rate of 7 ft/s, while the width is decreasing at the rate of 3 ft/s. At one time, the length is 12 feet and the diagonal is 13 feet. At this time find the rate of change in the perimeter.

A) 2 ft/s   B) 4 ft/s   C) 6 ft/s   D) 8 ft/s

E) 10 ft/s   F) 12 ft/s   G) 14 ft/s   H) 16 ft/s

Answer: 8 ft/s   medium

14. A particle starts at the origin and moves along the parabola $y = x^2$ such that its distance from the origin increases at 4 units per second. How fast is its x-coordinate changing as it passes through the point $(1, 1)$?

A) $4\sqrt{2}/3$ units/s   B) $\sqrt{2}/3$ units/s   C) $4\sqrt{2}$ units/s

D) $\sqrt{2}$ units/s   E) $4\sqrt{3}/3$ units/s   F) $\sqrt{3}/3$ units/s

G) $3\sqrt{3}$ units/s   H) $3\sqrt{3}/4$ units/s

Answer: $4\sqrt{2}/3$ units/s   medium

15. When a stone is dropped in a pool, a circular wave moves out from the point of impact at a rate of six inches per second. How fast is the area enclosed by the wave increasing when the wave is two inches in radius?

A) 12 in^2/s   B) 14 in^2/s   C) 16 in^2/s   D) 18 in^2/s

E) 20 in^2/s   F) 22 in^2/s   G) 24 in^2/s   H) 26 in^2/s

Answer: 24 in^2/s   medium

16. A particle moves along a path described by $y = x^2$. At what point along the curve are x and y changing at the same rate?

A) $(1, 1)$ B) $(1/2, 1/4)$ C) $(0, 0)$ D) $(1/3, 1/9)$

E) $(2/3, 4/9)$ F) $(1/4, 1/16)$ G) $(3/2, 9/4)$ H) $(4/3, 16/9)$

Answer: $(1/2, 1/4)$ medium

17. A mothball shrinks in such a way that its radius decreases by 1/6th inch per month. How fast is the volume changing when the radius is 1/4th inch? Assume the mothball is spherical.

A) $-\pi/12$ in^3/month   B) $-\pi/14$ in^3/month   C) $-\pi/16$ in^3/month

D) $-\pi/18$ in^3/month   E) $-\pi/20$ in^3/month   F) $-\pi/22$ in^3/month

G) $-\pi/24$ in^3/month   H) $-\pi/26$ in^3/month

Answer: $-\pi/24$ in^3/month medium

18. The electric resistance of a certain resistor as a function of temperature is given by $R = 6.000 + 0.002t^2$, where R is measured in Ohms and t in degrees Celsius. If the temperature is decreasing at the rate of 0.2°C per second, find the rate of change of resistance when $t = 38$°C.

A) -0.0304 ohms/s   B) -0.0204 ohms/s   C) -0.0104 ohms/s

D) -0.0404 ohms/s   E) 0.0104 ohms/s   F) 0.0204 ohms/s

G) 0.0304 ohms/s   H) 0.0404 ohms/s

Answer: -0.0304 ohms/s medium

Calculus, 2nd Edition
by James Stewart
Chapter 2, Section 9
Differentials and Linear Approximations

1.   Let $y = x^2$, $x = 2$, and $\Delta x = 1$. Find the value of the differential dy.

A) 2   B) 1/2   C) 1/3   D) 4

E) 1/4   F) 1/8   G) 3   H) 1

Answer: 4 easy

_____

2.   Let $y = x^2$, $x = 3$, and $\Delta x = 1$. Find the value of the corresponding change Δy.

A) 4   B) 8   C) 7   D) 2

E) 3   F) 6   G) 1   H) 5

Answer: 7 easy

_____

3.   Let $y = x^2$, $x = 1$, and $\Delta x = 0.1$. Find the error $dy - \Delta y$ in approximating Δy by dy.

A) −.2   B) −.1   C) −.03   D) −.07

E) −.05   F) −.01   G) −.02   H) −.3

Answer: −.01 medium

_____

4.   Use differentials to approximate $\sqrt{26}$.

A) 5.1   B) 5.2   C) 5.15   D) 5.3

E) 5.4   F) 5.25   G) 5.35   H) 5.05

Answer: 5.1 medium

_____

5.   The radius of a circle is given as 10 cm, with a possible error of measurement equal to 1 mm.
Use differentials to estimate the maximum error in the area, in cm^2.

A) 10π B) 2π C) 3π D) π

E) 8π F) 5π G) 6π H) 4π

Answer: 2π medium

_____

6. A spherical tank has radius equal to 10 feet (120 inches). Use differentials to estimate, in cubic inches, the amount of paint needed to cover the surface with a layer 1/100 of an inch thick.

A) 288π B) 1728π C) 144π D) 480π

E) 576π F) 512π G) 960π H) 640π

Answer: 576π medium

_____

7. Use differentials to approximate $\sqrt[5]{31}$.

A) 161/80   B) 77/40   C) 163/80   D) 81/40

E) 79/40   F) 159/80   G) 83/40   H) 157/80

Answer: 159/80   medium

_____

8. Use differentials to obtain an approximation for $\sqrt{16.2}$.

A) 4.026   B) 4.03   C) 4.025   D) 4.05

E) 4.015   F) .02498   G) 4.0185   H) 4.0245

Answer: 4.025   medium

_____

9. Let $y = x^4 + x^2 + 1$, $x = 1$, and $dx = 1$. Find the value of the differential dy.

A) 2   B) 4   C) 6   D) 8

E) 10   F) 12   G) 0   H) 1/2

Answer: 6   easy

_____

10. Let $y = 2x^3 + 3x - 4$, $x = 3$, and $dx = 1$. Find the value of the differential dy.

A) 21   B) 57   C) 165   D) 18

E) 54   F) 162   G) 59   H) 53

Answer: 57   easy

_____

11. Find the linearization of the function $f(x) = \sqrt{x+3}$ at $x_1 = 1$ and use it to approximate $\sqrt{3.98}$.

A) 2.005   B) 2.000   C) 1.995   D) 1.990

E) 1.985   F) 1.980   G) 1.975   H) 1.970

Answer: 1.995   medium

_____

12. Find the linearization of the function $f(x) = \sqrt{x+3}$ at $x_1 = 1$ and use it to approximate $\sqrt{4.05}$.

A) 2.0125   B) 2.0120   C) 2.0115   D) 2.0110

E) 2.0105   F) 2.0100   G) 2.0130   H) 2.0135

Answer: 2.0125 medium

Calculus, 2nd Edition
by James Stewart
Chapter 2, Section 10
Newton's Method

1. Use Newton's method with the initial approximation $x_1 = 2$ to find x_2, the second approximation to a root of the equation $x^2 - 2 = 0$.

   A) 5/7   B) 7/4   C) 4/3   D) 9/16

   E) 11/16   F) 5/4   G) 13/16   H) 3/2

   Answer: 3/2   medium

2. Use Newton's method with the initial approximation $x_1 = 1.5$ to find x_2, the second approximation to a root of the equation $x^2 - 2 = 0$.

   A) 19/13   B) 20/13   C) 19/12   D) 23/15

   E) 26/15   F) 17/12   G) 23/14   H) 19/14

   Answer: 17/12   medium

3. Use Newton's method with the initial approximation $x_1 = 2$ to find x_2, the second approximation to a root of the equation $x^5 - 31 = 0$.

   A) 161/80   B) 159/80   C) 83/40   D) 79/40

   E) 163/80   F) 81/40   G) 157/80   H) 77/40

   Answer: 159/80   medium

4. Use Newton's method with the initial approximation $x_1 = 2$ to find x_2, the second approximation to a root of the equation $x^5 - 34 = 0$.

   A) 79/40   B) 81/40   C) 77/40   D) 161/80

   E) 83/40   F) 157/80   G) 163/80   H) 159/80

   Answer: 81/40   medium

5. If Newton's method is used to find the cube root of a number a with a first approximation x_1, find an expression for x_2.

A) $x_2 = x_1 + \dfrac{3x_1^2}{x_1^3 - a}$

B) $x_2 = \sqrt[3]{x_1} - \sqrt[3]{a}$

C) $x_2 = x_1 - \dfrac{1}{3} x_1^{2/3} \left(x_1^{1/3} - a \right)$

D) $x_2 = x_1 + \dfrac{x_1^3 + a}{3x_1^2}$

E) $x_2 = x_1 + \dfrac{3x_1^2}{x_1^3 + a}$

F) $x_2 = x_1 - \dfrac{x_1^3 - a}{3x_1^2}$

G) $x_2 = x_1^{1/3} - \dfrac{1}{3} x_1^{-2/3}$

H) $x_2 = \sqrt[3]{a} + \dfrac{3x_1^2}{x_1^3 - a}$

Answer: $x_2 = x_1 - \dfrac{x_1^3 - a}{3x_1^2}$ hard
_____

6. If Newton's method is used to solve $2x^3 + 2x + 1 = 0$ with first approximation $x_1 = -1$, what is the second approximation, x_2?

A) $-.500$ B) $-.525$ C) $-.550$ D) $-.575$

E) $-.600$ F) $-.625$ G) $-.650$ H) $-.675$

Answer: $-.625$ medium
_____

7. If Newton's method is used to solve $x^3 + x + 1 = 0$ with an initial approximation $x_1 = -1$, what is the second approximation, x_2?

A) -0.75 B) -0.80 C) -0.85 D) -0.90

E) -0.95 F) -1.05 G) -1.10 H) -1.15

Answer: -0.75 medium
_____

8. If Newton's method is used to solve $x^3 + x^2 + 2 = 0$ with an initial approximation $x_1 = -2$, what is the second approximation, x_2?

A) -2.15 B) -2.10 C) -2.05 D) -1.95

E) -1.90 F) -1.85 G) -1.80 H) -1.75

Answer: -1.75 medium
_____

9. If Newton's method is used to find the square root of a number a with a first approximation x_1, find an expression for x_2.

A) $x_2 = x_1 + \dfrac{2x_2}{x_1^2 - a}$

B) $x_2 = \sqrt{x_1} - \sqrt{a}$

C) $x_2 = x_1 - \frac{1}{2}x_1(x_1^{1/2} - a)$

D) $x_2 = x_1 + \dfrac{x_1^2 + a}{2x_1^2}$

E) $x_2 = x_1 + \dfrac{2x_1^2}{x_1^2 + a}$

F) $x_2 = x_1 - \dfrac{x_1^2 - a}{2x_1}$

G) $x_2 = x_1^{1/2} - \frac{1}{2}x_1^{-1}$

H) $x_2 = \sqrt{a} + \dfrac{2x_1}{x_1^2 - a}$

Answer: $x_2 = x_1 - \dfrac{x_1^2 - a}{2x_1}$ hard

10. Use Newton's Method to find the root of $6x^3 + x^2 - 19x + 6 = 0$ that lies between 0 and 1.

A) 0.316   B) 0.333   C) 0.158   D) 0.167

E) 0.474   F) 0.500   G) 0.079   H) 0.084

Answer: 0.333   medium

11. Use Newton's Method to determine a root (to two decimal places) of the equation $x^3 + 8x - 23 = 0$, given an initial starting value of $x_1 = 2$.

A) 1.75   B) 1.80   C) 1.85   D) 1.90

E) 1.95   F) 2.00   G) 2.05   H) 2.10

Answer: 1.95   medium

12. Use Newton's Method to approximate a solution to the following equation: $x^3 + 2x = 3.1$.

A) 1.00   B) 1.01   C) 1.02   D) 1.03

E) 1.04   F) 1.05   G) 1.06   H) 1.07

Answer: 1.02   medium

Calculus, 2nd Edition
by James Stewart
Chapter 3, Section 1
Maximum and Minimum Values

1.   1. Find the minimum value of the function $f(x) = x^2 - 1/2$.

   A) -1 B) $3/4$ C) $1/2$ D) 1

   E) $-1/2$ F) 0 G) $-1/4$ H) $1/4$

   Answer: $-1/2$ easy
   _____

2.   Find the value x at which the minimum of the function $f(x) = x^2 - 1/2$ occurs.

   A) 1 B) $-1/4$ C) $-1/2$ D) $3/4$

   E) 0 F) $1/4$ G) -1 H) $1/2$

   Answer: 0 easy
   _____

3.   Find the minimum value of the function $f(x) = x^2 - x$.

   A) $-1/2$ B) $3/4$ C) $1/2$ D) 0

   E) $1/4$ F) 1 G) -1 H) $-1/4$

   Answer: $-1/4$ hard (because most answer $1/2$)
   _____

4.   Find the value x at which the minimum of the function $f(x) = x^2 - x$ occurs.

   A) $1/4$ B) 1 C) $-1/2$ D) 0

   E) $-1/4$ F) $3/4$ G) -1 H) $1/2$

   Answer: $1/2$ medium
   _____

5.   Find the distance between the two critical numbers of the function $f(x) = x^3 - 3x + 27$.

   A) 4 B) 1 C) 8 D) 3

   E) 2 F) 9 G) 6 H) 5

   Answer: 2 medium
   _____

6. Find the difference between the local maximum and the local minimum of the function $f(x) = x^3 - 3x + 27$.

A) 4   B) 1   C) 9   D) 2

E) 6   F) 5   G) 8   H) 3

Answer: 4 medium

_____

7. Find the absolute maximum of the function $f(x) = x^3 - x^2 - x$ on the interval $-10 \le x \le 1$.

A) $-5/27$ B) $-2/9$ C) $5/27$ D) $4/9$

E) $-7/27$ F) $7/27$ G) $-4/9$ H) $2/9$

Answer: $5/27$ medium

_____

8. Find the absolute maximum of the function $f(x) = x^3 - x^2 - x$ on the interval $-10 \le x \le 2$.

A) $1/9$ B) $1/5$ C) 0   D) $1/7$

E) 1   F) $1/4$ G) $1/3$ H) 2

Answer: 2 medium

_____

9. Find the absolute minimum and maximum values of the function $f(x) = 2x^3 - 3x^2 - 12x + 45$ on the closed interval $[-3, 3]$.

A) $41, 45$ B) $0, 52$ C) $25, 36$ D) $0, 36$

E) $0, 25$ F) $-41, 52$ G) $25, 52$ H) $-25, 52$

Answer: $0, 52$ medium

_____

10. Find the absolute minimum and maximum values of the function $f(x) = x^3 - x^2 + 11$ on the closed interval $[-3, 0.5]$.

A) $-3, 15$ B) $-25, 11$ C) $0, 17$ D) $-10, 30$

E) $-41, 20$ F) $-30, 40$ G) $-25, 87/8$ H) $-10, 20$

Answer: $-25, 11$ medium

_____

11. Find the absolute minimum and maximum values of the function $f(x) = 4x^3 - 15x^2 + 12x + 7$ on the closed interval $[0, 3]$.

A) $0, 3$ B) $0, 5$ C) $3, 5$ D) $3, 9.75$

E) $3, 16$ F) $5, 7$ G) $7, 16$ H) $5, 10.25$

Answer: $3, 16$ medium

12. Find the minimum and maximum values of $y = x^3 - 9x + 8$ on the interval $[-3, 1]$, if they exist.

A) $8, 8 + 6\sqrt{3}$ B) $0, 8 + 6\sqrt{3}$ C) $0, 8$

D) $8, 8 + \sqrt{3}$ E) $0, 8 + \sqrt{3}$ F) $8 + \sqrt{3}, 8 + 6\sqrt{3}$

G) $\sqrt{3}, 8$ H) $\sqrt{3}, 8 + 6\sqrt{3}$

Answer: $0, 8 + 6\sqrt{3}$ medium

13. Let $f(x) = x^{1/2}(1 - x)$ for $x \geq 0$. Find the absolute maximum of $f(x)$ on the interval $[0, 4]$.

A) 0 B) $2/(3\sqrt{3})$ C) -6 D) $1/3$

E) 4 F) $2/\sqrt{3}$ G) $\sqrt{3}$ H) $2\sqrt{3}$

Answer: $2/(3\sqrt{3})$ medium

14. Given that $f(x) = x^3 + ax^2 + bx$ has critical numbers at $x = 1$ and $x = 3$ find a and b.

A) $-9, 6$ B) $-8, 7$ C) $-7, 8$ D) $-6, 9$

E) $-9, 3$ F) $-8, 4$ G) $-7, 5$ H) $-6, 6$

Answer: $-6, 9$ hard

15. For the function $f(x)$, find the maximum and minimum values on the interval $[-1, 5]$

$$f(x) = \begin{cases} x^2 - 4 & \text{if } x \leq 2 \\ x^2 - 8x + 12 & \text{if } x > 2 \end{cases}$$

A) $0, -4$ B) $4, 0$ C) $2, 0$ D) $4, -4$

E) $2, -4$ F) $4, -2$ G) $2, -2$ H) $0, -2$

Answer: $0, -4$ medium

Calculus, 2nd Edition
by James Stewart
Chapter 3, Section 2
The Mean Value Theorem

1. According to Rolle's Theorem, what is the largest number of real roots that the equation $x^7 + x = 0$ can have?

   A) 4   B) 5   C) 1   D) 0

   E) 2   F) 7   G) 3   H) 6

   Answer: 1 medium

2. According to Rolle's Theorem, what is the largest number of real roots that the equation $x^7 - x = 0$ can have?

   A) 2   B) 5   C) 7   D) 6

   E) 4   F) 1   G) 0   H) 3

   Answer: 3 medium

3. How many real roots does the equation $x^7 + x = 0$ have?

   A) 2   B) 7   C) 5   D) 1

   E) 6   F) 4   G) 0   H) 3

   Answer: 1 medium

4. How many real roots does the equation $x^7 - x = 0$ have?

   A) 4   B) 3   C) 1   D) 2

   E) 6   F) 5   G) 7   H) 0

   Answer: 3 medium

5. According to Rolle's Theorem, what is the largest number of real roots that the equation $x^7 - x + 17 = 0$ can have?

   A) 2   B) 1   C) 3   D) 0

   E) 7   F) 5   G) 6   H) 4

   Answer: 3 medium

6. How many real roots does the equation $x^7 - x + 17 = 0$ have?

   A) 7 B) 5 C) 1 D) 3

   E) 2 F) 6 G) 0 H) 4

   Answer: 1 hard
   _____

7. Consider the function $f(x) = x^2$ on the interval $[0, 1/2]$. According to the mean value theorem, there must be a number c in $(0, 1/2)$ such that $f'(c)$ is equal to a particular value d. What is d?

   A) 3/2 B) 1 C) 1/2 D) 2

   E) 2/3 F) 3/4 G) 1/4 H) 3

   Answer: 1/2 medium
   _____

8. Tell which of the statements below are true:

   1) If $f'(c) = 0$, then $f(x)$ has a maximum or minimum value at $x = c$.
   2) If $f'(x) = g'(x)$ for all x in an interval I, then $f(x) = g(x)$ on I.
   3) If $f(x)$ is differentiable on the open interval (a,b), and c is a point of local maximum for f in (a, b), then $f'(c) = 0$.

   A) none B) 1 C) 2 D) 3

   E) 1, 2 F) 1, 3 G) 2, 3 H) all

   Answer: 3 medium
   _____

9. What value(s) of c (if any) are predicted by the Mean Value Theorem for the function $f(x) = (x - 2)^3$ on the interval $[0, 2]$?

   A) 4

   B) $2 + \dfrac{2}{\sqrt{3}}$

   C) $2 - \dfrac{2}{\sqrt{3}}$

   D) 0

   E) $2 + \dfrac{2}{\sqrt{3}}, \ 2 - \dfrac{2}{\sqrt{3}}$

   F) $4, \ 2 + \dfrac{2}{\sqrt{3}}, \ 2 - \dfrac{2}{\sqrt{3}}$

   G) $0, \ 2 - \dfrac{2}{\sqrt{3}}$

   H) no values predicted

   Answer: $2 - \dfrac{2}{\sqrt{3}}$ medium
   _____

Calculus, 2nd Edition
by James Stewart
Chapter 3, Section 3
Monotonic Functions and the First Derivative Test

1. At what value of x does the function $f(x) = x^2 - 2x$ change from decreasing to increasing?

   A) 1   B) -1 C) $-1/2$ D) 2

   E) $1/2$ F) -2 G) 0   H) $-3/2$

   Answer: 1   easy

2. Find the length of the largest interval on which the function $f(x) = x - \sqrt{x}$ is decreasing.

   A) $3/2$ B) $\sqrt{2}$ C) $\sqrt{3}/4$ D) $\sqrt{2}/2$

   E) $\sqrt{3}/2$ F) $1/2$ G) $2/3$ H) $1/4$

   Answer: $1/4$ hard

3. At what value of x does the function $f(x) = 3x - \sqrt[3]{x}$ change from decreasing to increasing?

   A) $1/8$ B) $1/2$ C) $1/3$ D) $1/27$

   E) $1/9$ F) $\sqrt{3}/4$ G) $\sqrt{3}/3$ H) $1/32$

   Answer: $1/27$ hard

4. At what value of x does the function $f(x) = x^3 - 3x^2 - 9x$ change from increasing to decreasing?

   A) 4   B) 3   C) -2 D) -1

   E) 0   F) 2   G) 1   H) -3

   Answer: -1 medium

5. At what value of x does the function $f(x) = x^3 - 3x^2 - 9x$ change from decreasing to increasing?

   A) 3   B) 1   C) 4   D) -3

   E) 2   F) 0   G) -2 H) -1

   Answer: 3   medium

6. What is the length of the largest interval on which the function $f(x) = x^3 - 3x^2 - 9x$ is decreasing?

    A) $3\sqrt{2}$ B) 1   C) 4   D) $2\sqrt{2}$

    E) 3   F) $2\sqrt{3}$ G) $\sqrt{3}$ H) 2

    Answer: 4 medium
    _____

7. What is the length of the largest interval on which the function $\dfrac{x}{x^2+1}$ is increasing?

    A) 3   B) $2\sqrt{3}$ C) $2\sqrt{2}$ D) 1

    E) 4   F) $\sqrt{3}$ G) 2   H) $3\sqrt{2}$

    Answer: 2 hard
    _____

8. The function $f(x) = x + \dfrac{1}{x}$, $x \neq 0$, is

    A) increasing on $(-1, 0)$ B) decreasing on $(0, 1)$

    C) increasing on $(0, 1)$ D) decreasing on $(1, \infty)$

    E) increasing on $(0, \infty)$ F) decreasing on $(0, \infty)$

    G) increasing on $(-\infty, 0)$ H) decreasing on $(-\infty, 0)$

    Answer: decreasing on $(0, 1)$ (medium)
    _____

9. On what interval is the function $f(x) = \dfrac{x}{x^2+1}$ increasing?

    A) $[-1, 1]$ B) $[-1, 2]$ C) $[-2, 1]$ D) $[-2, 2]$

    E) $(-\infty, 1]$ F) $(-\infty, 2)$ G) $[-2, \infty)$ H) $(-\infty, \infty)$

    Answer: $[-1, 1]$ medium
    _____

10. Find the interval on which $f(x) = x - 2\sin x$, $0 \leq x \leq 2\pi$, is increasing.

    A) $[\pi/3, 5\pi/3]$ B) $[0, \pi/3]$ C) $[5\pi/3, 2\pi]$ D) $[0, \pi/2]$

    E) $[\pi/2, 3\pi/2]$ F) $[3\pi/2, 2\pi]$ G) $[0, \pi]$ H) $[\pi, 2\pi]$

    Answer: $[\pi/3, 5\pi/3]$ medium
    _____

11. Find the interval on which $f(x) = x + \cos x$, $0 \leq x \leq 2\pi$, is increasing.

A) $[\pi/3, 5\pi/3]$ B) $[0, \pi/3]$ C) $[5\pi/3, 2\pi]$ D) $[0, \pi/2]$

E) $[\pi/2, 3\pi/2]$ F) $[3\pi/2, 2\pi]$ G) $[0, \pi]$ H) $[0, 2\pi]$

Answer: $[0, 2\pi]$ medium
_____

12. Find the interval on which $f(x) = x \sin x + \cos x$, $0 \leq x \leq \pi$, is increasing.

A) $[\pi/3, 5\pi/3]$ B) $[0, \pi/3]$ C) $[5\pi/3, 2\pi]$ D) $[0, \pi/2]$

E) $[\pi/2, 3\pi/2]$ F) $[3\pi/2, 2\pi]$ G) $[0, \pi]$ H) $[\pi, 2\pi]$

Answer: $[0, \pi/2]$ medium
_____

13. Find the intervals on which $f(x) = 2\tan x - \tan^2 x$ is increasing.

A) $[n\pi - \pi/2, n\pi + \pi/4]$, n an integer   B) $[n\pi + \pi/2, n\pi + 3\pi/2]$, n an integer

C) $[n\pi - \pi/3, n\pi]$, n an integer   D) $[n\pi, n\pi + \pi/2]$, n an integer

E) $[n\pi - \pi/2, n\pi + \pi/2]$, n an integer   F) $[n\pi - \pi/3, n\pi + \pi/2]$, n an integer

G) $[n\pi, n\pi + \pi/3]$, n an integer   H) $[n\pi - 2\pi/3, n\pi]$, n an integer

Answer: $[n\pi - \pi/2, n\pi + \pi/4]$ hard
_____

Calculus, 2nd Edition
by James Stewart
Chapter 3, Section 4
Concavity and Points of Inflection

1. Find the x-coordinate of the point of inflection of the function $x^3 - x^2 - x + 1$.

    A) $-3/4$ B) $-3/2$ C) $3/4$ D) $-1/3$
    E) $3/2$ F) $1/3$ G) $-2/3$ H) $2/3$

    Answer: $1/3$ medium
    _____

2. Find the y-coordinate of the point of inflection of the function $x^3 - x^2$.

    A) $1/3$ B) $-2/27$ C) $2/27$ D) $-1/3$
    E) $-2/9$ F) $2/9$ G) $-2/3$ H) $2/3$

    Answer: $-2/27$ medium
    _____

3. Find the length of the largest interval on which the function $x^4 - 6x^3$ is concave down.

    A) $\sqrt{2}/2$ B) 3 C) $2\sqrt{2}$ D) $\sqrt{2}$
    E) 1 F) 4 G) $\sqrt{2}/3$ H) 2

    Answer: 3 medium
    _____

4. Find the length of the largest interval on which the function $f(x) = x^4 - 12x^3 + 3x - 10$ is
    concave down.

    A) $\sqrt{2}/2$ B) 6 C) $\sqrt{2}/3$ D) $2\sqrt{2}$
    E) 4 F) $\sqrt{2}$ G) 3 H) 2

    Answer: 6 medium
    _____

5. How many points of inflection does the function $f(x) = x^7 - x^2$ have?

    A) 5 B) 0 C) 2 D) 7
    E) 6 F) 1 G) 3 H) 4

    Answer: 1 medium
    _____

6. How many points of inflection does the function $f(x) = x^8 - x^2$ have?

A) 1   B) 0   C) 4   D) 5

E) 6   F) 3   G) 2   H) 7

Answer: 2 medium

7. How many points of inflection does the function $f(x) = x^8 + x^2$ have?

A) 7   B) 3   C) 2   D) 6

E) 0   F) 1   G) 5   H) 4

Answer: 0 medium

8. On what interval is the graph of $f(x) = \left[1 - \frac{1}{x}\right]^2$ concave downward?

A) $(-\infty, 0)$ B) $(3/2, \infty)$ C) $(0, \infty)$ D) $(1, \infty)$

E) $(-\infty, -1)$ F) $(1, 3/2)$ G) $(-3/2, -1)$ H) $(0, 1)$

Answer: $(3/2, \infty)$ medium

9. Determine the largest interval on which the function $f(x) = x + \frac{1}{x}$ is concave upward.

A) $(-1, 0)$ B) $(-\infty, -1)$ C) $(-\infty, 0)$ D) $(0, \infty)$

E) $(1, \infty)$ F) $(0, 1)$ G) $(0, 1/2)$ H) $(-1/2, 0)$

Answer: $(0, \infty)$ medium

10. Find the largest interval on which the function $f(x) = \frac{x}{x^2 + 1}$ is concave upward.

A) $(0, 1)$ B) $(1, 2)$ C) $(1, \infty)$ D) $(0, \infty)$

E) $(1, \sqrt{3})$ F) $(\sqrt{3}, \infty)$ G) $(\sqrt{2}, \infty)$ H) $(1/2, \infty)$

Answer: $(\sqrt{3}, \infty)$ hard

11. Determine α so that the function $f(x) = x^2 + \frac{\alpha}{x}$ has an inflection point at $x = 1$.

A) -3 B) -2 C) -1 D) 0

E) 1   F) 2   G) 3   H) 4

Answer: -1 medium

12. Find the values α, β, γ so the function $f(x) = x^3 + \alpha x^2 + \beta x + \gamma$ has a critical point at $(1, 5)$ and an inflection point at $(2, 3)$. All listed answers are in the form α, β, γ.

A) $-6, 9, 1$ B) $1, -6, 9$ C) $9, 1, -6$ D) $1, 9, -6$

E) $-3, 7, -1$ F) $-1, -3, 7$ G) $7, -1, -3$ H) $-1, 7, -3$

Answer: $-6, 9, 1$ medium
_____

13. Find the x-coordinates of the points of inflection of $f(\theta) = \sin^2 \theta$.

A) $n\pi + \pi/4$, n an integer   B) $n\pi - \pi/4$, n an integer

C) $n\pi \pm \pi/4$, n an integer   D) $n\pi + \pi/3$, n an integer

E) $n\pi - \pi/3$, n an integer   F) $n\pi \pm \pi/3$, n an integer

G) $n\pi$, n an integer   H) $n\pi/2$, n an integer

Answer: $n\pi \pm \pi/4$, n an integer   medium
_____

14. Find the x-coordinates of the points of inflection of $f(\theta) = \theta + \sin \theta$.

A) $n\pi + \pi/4$, n an integer   B) $n\pi - \pi/4$, n an integer

C) $n\pi \pm \pi/4$, n an integer   D) $n\pi + \pi/3$, n an integer

E) $n\pi - \pi/3$, n an integer   F) $n\pi \pm \pi/3$, n an integer

G) $n\pi$, n an integer   H) $n\pi/2$, n an integer

Answer: $n\pi$, n an integer   medium
_____

15. Find the intervals on which $f(\theta) = \cos^2 \theta$ is concave up.

A) $(n\pi + \pi/4, n\pi + 3\pi/4)$, n an integer   B) $(n\pi - \pi/4, n\pi + \pi/4)$, n an integer

C) $(n\pi + \pi/4, n\pi)$, n an integer   D) $(n\pi + \pi/3, n\pi + 2\pi/3)$, n an integer

E) $(n\pi - \pi/3, n\pi + \pi/3)$, n an integer   F) $(n\pi, n\pi + \pi/3)$, n an integer

G) $(n\pi, n\pi + 2\pi/3)$, n an integer   H) $(n\pi/2, n\pi)$, n an integer

Answer: $(n\pi + \pi/4, n\pi + 3\pi/4)$, n an integer   medium
_____

16. Find the intervals on which $f(\theta) = \cos^2 \theta$ is concave down.

A) $\left(n\pi + \pi/4,\ n\pi + 3\pi/4\right)$, n an integer
B) $\left(n\pi - \pi/4,\ n\pi + \pi/4\right)$, n an integer
C) $\left(n\pi + \pi/4,\ n\pi\right)$, n an integer
D) $\left(n\pi + \pi/3,\ n\pi + 2\pi/3\right)$, n an integer
E) $\left(n\pi - \pi/3,\ n\pi + \pi/3\right)$, n an integer
F) $\left(n\pi,\ n\pi + \pi/3\right)$, n an integer
G) $\left(n\pi,\ n\pi + 2\pi/3\right)$, n an integer
H) $\left(n\pi/2,\ n\pi\right)$, n an integer

Answer: $\left(n\pi - \pi/4,\ n\pi + \pi/4\right)$, n an integer medium

17. Find the intervals on which $f(\theta) = \theta + \cos \theta$ is concave up.

A) $\left(2n\pi + \pi/4,\ 2n\pi + 3\pi/4\right)$, n an integer
B) $\left(2n\pi - \pi/4,\ 2n\pi + \pi/4\right)$, n an integer
C) $\left(2n\pi + \pi/4,\ 2n\pi\right)$, n an integer
D) $\left(2n\pi + \pi/3,\ 2n\pi + 2\pi/3\right)$, n an integer
E) $\left(2n\pi - \pi/3,\ 2n\pi + \pi/3\right)$, n an integer
F) $\left(2n\pi - \pi/2,\ 2n\pi + \pi/2\right)$, n an integer
G) $\left(2n\pi - \pi/2,\ 2n\pi + 3\pi/2\right)$, n an integer
H) $\left(2n\pi + \pi/2,\ 2n\pi + 3\pi/2\right)$, n an integer

Answer: $\left(2n\pi + \pi/2,\ 2n\pi + 3\pi/2\right)$, n an integer medium

18. Find the intervals on which $f(\theta) = \theta + \cos \theta$ is concave down.

A) $\left(2n\pi + \pi/4,\ 2n\pi + 3\pi/4\right)$, n an integer
B) $\left(2n\pi - \pi/4,\ 2n\pi + \pi/4\right)$, n an integer
C) $\left(2n\pi + \pi/4,\ 2n\pi\right)$, n an integer
D) $\left(2n\pi + \pi/3,\ 2n\pi + 2\pi/3\right)$, n an integer
E) $\left(2n\pi - \pi/3,\ 2n\pi + \pi/3\right)$, n an integer
F) $\left(2n\pi - \pi/2,\ 2n\pi + \pi/2\right)$, n an integer
G) $\left(2n\pi - \pi/2,\ 2n\pi + 3\pi/2\right)$, n an integer
H) $\left(2n\pi + \pi/2,\ 2n\pi + 3\pi/2\right)$, n an integer

Answer: $\left(2n\pi - \pi/2,\ 2n\pi + \pi/2\right)$, n an integer medium

Calculus, 2nd Edition
by James Stewart
Chapter 3, Section 5
Limits at Infinity; Horizontal Asymptotes

1. Find the value of the limit $\lim\limits_{x \to \infty} \dfrac{7 + 3x}{4 - x}$.

   A) 3/4   B) 3   C) 7   D) −3/4

   E) −3   F) −7/4   G) −7   H) 7/4

   Answer: −3 easy
   _____

2. Find the value of the limit $\lim\limits_{x \to \infty} 3x^{-1/2}$.

   A) 1/3   B) $-\sqrt{2}/2$ C) $\sqrt{2}$ D) $\sqrt{2}$

   E) $\sqrt{2}/2$ F) 0   G) 3   H) $\sqrt{3}$

   Answer: 0 easy
   _____

3. Find the value of the limit $\lim\limits_{x \to \infty} \dfrac{x^2 - 1}{x^2 + 2x + 1}$.

   A) −1/2   B) 1   C) 2   D) 1/2

   E) −1   F) −1/3   G) −2   H) 1/3

   Answer: 1 medium
   _____

4. Find the value of the limit $\lim\limits_{x \to \infty} \sqrt{\dfrac{x + 8x^2}{2x^2 - 1}}$.

   A) −2   B) 1   C) −1/2   D) 4

   E) 1/2   F) 2   G) 0   H) 1

   Answer: 2 medium
   _____

5. Find the value of the limit $\lim\limits_{x \to \infty} \dfrac{\sin^3 x}{x}$.

   A) −1/3   B) 1   C) 1/3   D) $-\sqrt{2}/2$

   E) −1   F) $\sqrt{2}/2$ G) 0   H) 3

   Answer: 0 medium
   _____

6.	Find the value of the limit $\lim_{x \to \infty} \left(\sqrt{x^2 + x + 1} - x \right)$.

A) $-1/2$	B) $\sqrt{2}$	C) $-\sqrt{2}$	D) $1/4$

E) $\sqrt{2}/2$	F) $-1/4$	G) $-\sqrt{2}/2$	H) $1/2$

Answer: $1/2$ hard

_____

7.	Find the value of the limit $\lim_{x \to \infty} \cos \frac{1}{x}$.

A) $\sqrt{2}/2$	B) $-1/2$	C) $-\sqrt{3}/2$	D) 1

E) $\sqrt{3}/2$	F) 0	G) -1	H) $-\sqrt{2}/2$

Answer: 1 medium

_____

8.	Find the value of the limit $\lim_{x \to \infty} \left[\sqrt{x^2 + x} - \sqrt{x^2 - x} \right]$.

A) $-\infty$	B) -2	C) -1	D) 0

E) $1/2$	F) 1	G) $3/2$	H) ∞

Answer: 1 hard

_____

9.	Find the value of the limit $\lim_{x \to \infty} \dfrac{6x^4 - 8x^2 + 3x - 4}{7 + 6x^3 - 2x^4}$.

A) ∞	B) $-\infty$	C) 1	D) -1

E) 3	F) -3	G) 0	H) does not exist

Answer: -3 medium

_____

10.	Find the value of the limit $\lim_{x \to \infty} \dfrac{3x^2 + 5x - 1}{2x^2 - 5x + 1}$.

A) $-\infty$	B) -2	C) -1	D) 0

E) $1/2$	F) 1	G) $3/2$	H) ∞

Answer: $3/2$ medium

_____

11.	Find the value of the limit $\lim_{x \to -\infty} \dfrac{\sqrt{x^2 + 4x}}{4x + 1}$.

A) 0	B) 4	C) -4	D) $1/4$

E) $-1/4$	F) ∞	G) $-\infty$	H) does not exist

Answer: $-1/4$ medium

_____

12. Find the value of the limit $\lim_{x \to \infty} \left(\sqrt{x^2 + 3x} - x\right)$.

A) 3 B) 3/2 C) 0 D) 1

E) 2 F) ∞ G) $-\infty$ H) does not exist

Answer: 3/2 hard

13. Find the value of the limit $\lim_{x \to \infty} \left(\sqrt{x^2 + 2x} - x\right)$.

A) $-\infty$ B) -2 C) -1 D) 0

E) 1 F) 2 G) ∞ H) does not exist

Answer: 1 hard

14. Find the value of the limit $\lim_{x \to \infty} \left(\sqrt{x^2 + x} - x\right)$.

A) $-\infty$ B) -2 C) -1 D) 0

E) 1/2 F) 1 G) 3/2 H) ∞

Answer: 1/2 hard

15. Find the horizontal asymptote of the curve $y = \dfrac{3x}{x+4}$.

A) $y = 3$ B) $y = 0$ C) $y = -3$ D) $y = 2$

E) $y = -1$ F) $y = 1$ G) $y = -2$ H) $y = -4$

Answer: $y = 3$ medium

16. Find the value of the limit $\lim_{x \to \infty} \dfrac{4x^2 + x}{x^2 - 2}$ if possible.

A) -2 B) 1 C) 2 D) 4

E) -1 F) -3 G) -4 H) 3

Answer: 4 medium

17. Find the limit of $x - \sqrt{x^2 - 4}$ as x approaches infinity.

A) -2 B) 1 C) 2 D) 4

E) -1 F) -3 G) 0 H) 3

Answer: 0 medium

18. Find the value of the limit $\displaystyle \lim_{x \to \infty} \frac{3x^2 + 4x + 2}{5 - x + x^3}$.

A) $-1/2$ B) 1   C) 2   D) $1/2$

E) -1 F) $-1/3$ G) -2 H) 0

Answer: 0 medium

Calculus, 2nd Edition
by James Stewart
Chapter 3, Section 6
Infinite Limits; Vertical Asymptotes

1. Find the value of the limit $\lim\limits_{x \to 0} \dfrac{4}{x^4}$.

   A) 1   B) 4   C) ∞ D) $-\infty$

   E) $-1/4$ F) $1/4$ G) 0   H) -4

   Answer: ∞ easy
   _____

2. Find the value of the limit $\lim\limits_{x \to 2} \dfrac{3}{(x-2)^2}$.

   A) ∞ B) $3/2$ C) $-3/2$ D) 3

   E) $3/4$ F) $-\infty$ G) -3 H) $-3/4$

   Answer: ∞ easy
   _____

3. Find the value of the limit $\lim\limits_{x \to 0} \dfrac{3}{(x-2)^2}$.

   A) $-3/2$ B) $-3/4$ C) $-\infty$ D) ∞

   E) $3/4$ F) -3 G) $3/2$ H) 3

   Answer: $3/4$ easy
   _____

4. Find the value of the limit $\lim\limits_{x \to -3^+} \dfrac{x}{x+3}$.

   A) -1 B) ∞ C) 3   D) 0

   E) 1   F) $-\infty$ G) -3 H) $1/3$

   Answer: $-\infty$ medium
   _____

5. Find the value of the limit $\lim\limits_{x \to -3^-} \dfrac{x+2}{x+3}$.

   A) $2/3$ B) 1   C) 0   D) ∞

   E) -3 F) $-\infty$ G) 3   H) $-2/3$

   Answer: ∞ medium
   _____

6. Find the value of the limit $\lim\limits_{x \to \infty} \dfrac{1}{\sqrt{x+1} - \sqrt{x}}$.

A) 1   B) 1/2   C) 2   D) −2

E) ∞   F) −∞   G) −1/2   H) 0

Answer: ∞   hard

7. Find the value of x at which the curve $y = \dfrac{3x}{x+4}$ has a vertical asymptote.

A) −1   B) 3   C) 2   D) −4

E) 1   F) 0   G) −2   H) −3

Answer: −4   medium

8. Find the value of the limit $\lim\limits_{x \to -3^+} \left(\dfrac{x^3}{x^2 + 5x + 6} \right)$.

A) 27/12   B) 3/2   C) −∞   D) ∞

E) 9/16   F) −3/2   G) 5/12   H) does not exist

Answer: ∞   medium

9. Find the value(s) of x at which the curve $y = \dfrac{x^2 - 4}{9 - x^2}$ has a vertical asymptote(s).

A) −3   B) 3   C) 2   D) −2

E) 3, 2   F) −3, −2   G) −3, −3, 2   H) −3, 3

Answer: −3, 3   medium

10. Find the value of x at which the curve $y = \dfrac{x^2 - 16}{x^2 - 5x + 4}$ has a vertical asymptote.

A) −1   B) 3   C) 2   D) −4

E) 1   F) 0   G) −2   H) −3

Answer: 1   medium

11. Find the value(s) of x at which the curve $y = \dfrac{3x^2 + 100}{4x^2 - 100}$ has a vertical asymptote(s).

A) −5   B) 5   C) −5, 5   D) −4

E) 4   F) −4, 4   G) −4, 4, 5   H) −5, −4, 4

Answer: −5, 5   medium

12. Find the value of x at which the curve $y = \dfrac{x^3 - 2x + 3}{x^2 + 4x + 4}$ has a vertical asymptote.

A) −1 B) 3 C) 2 D) −4

E) 1 F) 0 G) −2 H) −3

Answer: −2 medium

Calculus, 2nd Edition
by James Stewart
Chapter 3, Section 7
Curve Sketching

1. Find the smallest number in the domain of the function $f(x) = \sqrt{1 - x - x^2}$.

    A) $(-1 - \sqrt{5})/2$ B) $(1 + \sqrt{5})/2$ C) $(-1 + \sqrt{5})/2$

    D) $-\sqrt{3}/2$ E) $\sqrt{3}/2$ F) $-\sqrt{3}$

    G) $(1 - \sqrt{5})/2$ H) $\sqrt{3}$

    Answer: $(-1 - \sqrt{5})/2$ medium
    _____

2. Find the period of the function $f(x) = \tan 2x$.

    A) $3\pi/2$ B) $3\pi/4$ C) 3π D) $\pi/2$

    E) 4π F) 2π G) $\pi/3$ H) π

    Answer: $\pi/2$ medium
    _____

3. Find the slope of the slant asymptote of the curve $y = \dfrac{3x^2}{2x - 1}$.

    A) $3/2$ B) $-2/3$ C) $-1/3$ D) $1/2$

    E) $1/3$ F) $-1/2$ G) $2/3$ H) $-3/2$

    Answer: $3/2$ medium
    _____

4. Find the critical number c for $f(x) = x^4 - 3x^3 + 3x^2 - x$ at which $f(c)$ is not a local maximum and not a local minimum.

    A) $-1/2$ B) -3 C) -1 D) 0

    E) $-1/3$ F) 3 G) $1/3$ H) 1

    Answer: 1 hard
    _____

5. Find the distance along the x-axis between the point of local minimum and the point of local maximum of the function $f(x) = x^3 - 3x + 7$.

    A) $1/3$ B) 3 C) $3/7$ D) $49/3$

    E) 2 F) 1 G) $7/3$ H) $1/2$

    Answer: 2 medium
    _____

6. Find the value of x at which the direction of concavity of the function $f(x) = \frac{7x^2}{2x-3}$ changes.

A) 7/2   B) 3/2   C) 2/3   D) $-7/2$

E) $-3/2$ F) $-2/3$ G) 2/7   H) $-2/7$

Answer: 3/2 medium

_____

7. Find the maximum value of the function $f(x) = \frac{x^2 + 2x - 4}{x^2}$.

A) -5 B) -1 C) $-1/2$ D) 1/4

E) 11/9   F) 5/4   G) 7/4   H) 9/4

Answer: 5/4 medium

_____

8. Determine the largest interval on which the function $f(x) = \frac{x^2 + 2x - 4}{x^2}$ is concave upward.

A) $(-\infty, -4)$ B) $(-\infty, -2)$ C) $(-\infty, 0)$ D) $(-4, 0)$

E) $(0, 2)$ F) $(0, 4)$ G) $(4, \infty)$ H) $(6, \infty)$

Answer: $(6, \infty)$ medium

_____

9. On what interval is the function $f(x) = \frac{x^2 + 2x - 4}{x^2}$ increasing?

A) $(-\infty, -4]$ B) $(-\infty, -2]$ C) $(-\infty, 0)$ D) $[-4, -2]$

E) $[-4, 0)$ F) $(0, 4]$ G) $(0, \infty)$ H) $[4, \infty)$

Answer: $(0, 4]$ hard

_____

10. Sketch the curve $y = x^4 - 6x^2$.

A)

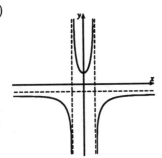

B)

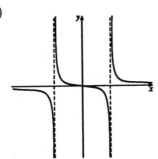

C)

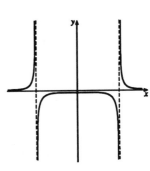

D)

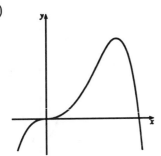

E)

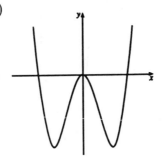

F)

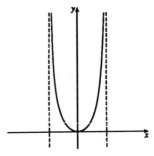

G)

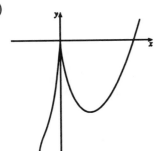

H)

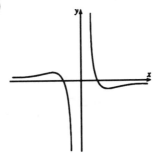

Answer: medium

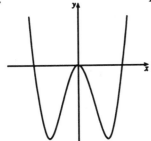

11. Sketch the curve $y = 4x^3 - x^4$.

A)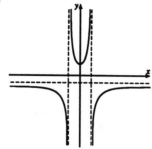

B)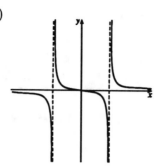

C)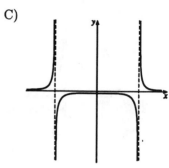

D)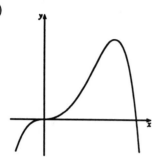

E)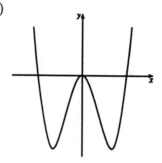

F)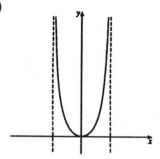

G)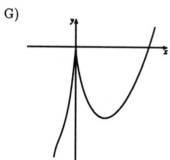

H)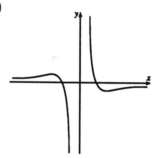

Answer: 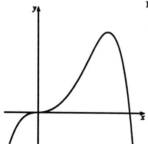 medium

12. Sketch the curve $y = \dfrac{1}{x^2 - 9}$.

A)

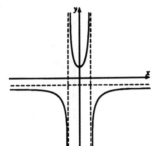

B)

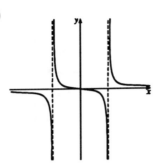

C)

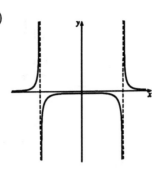

D)

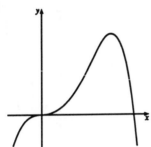

E)

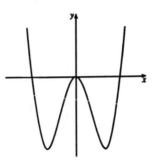

F)

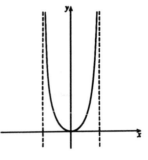

G)

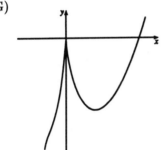

H)

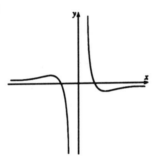

Answer:

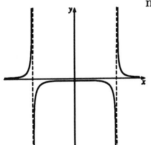

13. Sketch the curve $y = \dfrac{x}{x^2 - 9}$.

A)

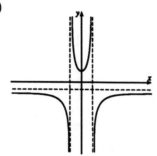

B)

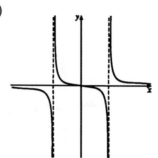

C)

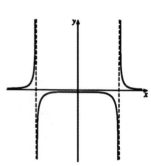

D)

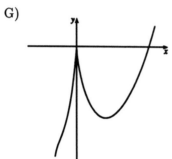

E)

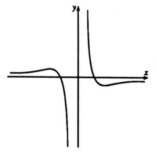

F)

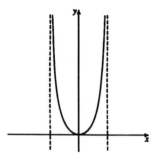

G)

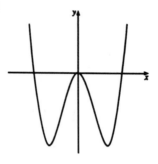

H)

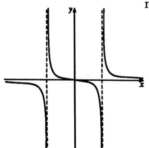

Answer: medium

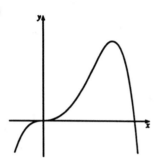

14. Sketch the curve $y = \dfrac{1+x^2}{1-x^2}$.

A)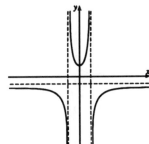

B)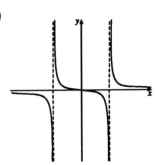

C)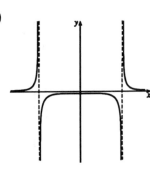

D)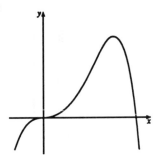

E)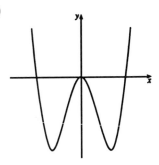

F)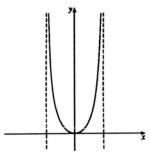

G)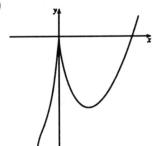

H)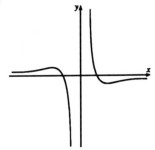

Answer: medium

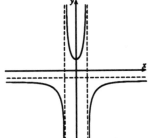

15. Sketch the curve $y = \dfrac{1-x^2}{x^3}$.

A)

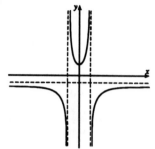

B)

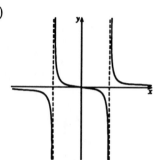

C)

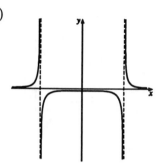

D)

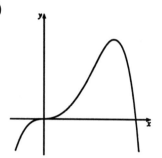

E)

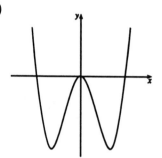

F)

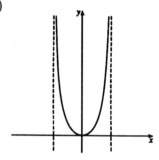

G)

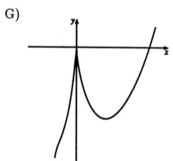

H)

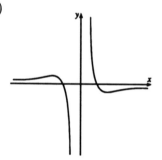

Answer: medium

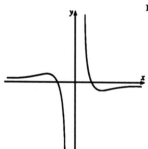

16. Sketch the curve $y = x^{5/3} - 5x^{2/3}$.

A)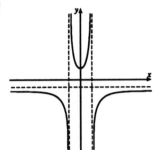

B)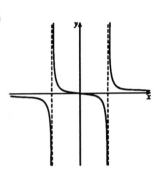

C)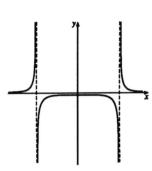

D)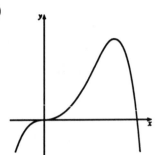

E)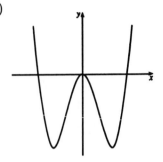

F)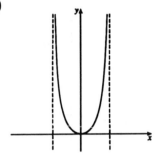

G)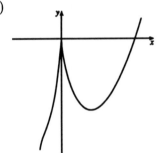

H)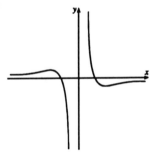

Answer: 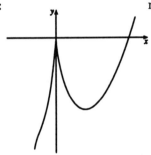 medium

17. Sketch the curve $y = 2x + \cot x$, $0 < x < \pi$.

A)

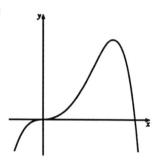

B)

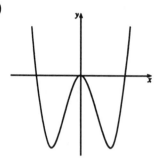

C)

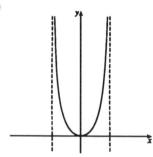

D)

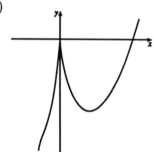

E)

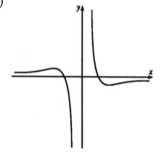

F)

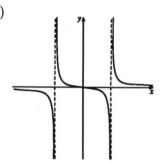

G)

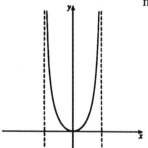

H)

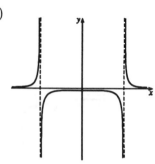

Answer: medium

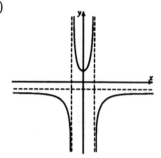

Calculus, 2nd Edition
by James Stewart
Chapter 3, Section 8
Applied Maximum and Minimum Problems

1. We are given two numbers that sum to 100, and we wish to make their product as large as possible. How large is that product?

    A) 2100   B) 2000   C) 3000   D) 2700

    E) 2250   F) 3200   G) 2500   H) 1850

    Answer: 2500 easy

    _____

2. Two numbers a and b sum to 60. We wish to make ab^2 as large as possible. Find the value of a.

    A) 30   B) 20   C) 18   D) 21

    E) 32   F) 16   G) 28   H) 24

    Answer: 20 medium

    _____

3. A farmer has 20 feet of fence, and he wishes to make from it a rectangular pen for his pig Wilbur, using a barn as one of the sides. In square feet, what is the maximum area possible for this pen?

    A) 64   B) 75   C) 60   D) 32

    E) 56   F) 25   G) 40   H) 50

    Answer: 50 easy

    _____

4. It is desired to make a rectangular pen with perimeter equal to 40 feet. It is not necessary, however, to achieve maximum area. In fact, any area that is greater than or equal to 75% of the maximum area will be satisfactory. Under these conditions, find the maximum number of feet the longer side can be.

    A) 15   B) 14   C) 12.5   D) 12

    E) 11   F) 13   G) 13.5   H) 10.5

    Answer: 15 medium

    _____

5. A lakefront runs east-west. A man in a rowboat is 5 miles due north of a point A on the shore. He wishes to get to B, 5 miles due east of A, in the least time. He rows 3 miles per hour and walks 5 miles per hour. What is the minimum time, in minutes?

   A) 180   B) 160   C) 140   D) 120
   E) 90   F) 150   G) 170   H) 130

   Answer: 140 hard
   _____

6. A drinking cup is made in the shape of a right circular cylinder. For a fixed volume, we wish to make the total material used, the circular bottom and the cylindrical side, as small as possible. Under this condition, what is the ratio of the height to the diameter?

   A) 4   B) $\pi/4$ C) 2   D) 1
   E) $\pi/8$ F) $4/\pi$ G) $8/\pi$ H) 1/2

   Answer: 1/2 hard
   _____

7. A right circular cylindrical container is made by cutting the circular top and bottom out of two squares whose sides are equal to the diameter of the top and bottom. The remnants of the squares are wasted. Therefore, subject to fixed volume, it is required to minimize the total area of the two squares plus that of the side of the container. What is the ratio of the height to the diameter?

   A) $4/\pi$ B) 2   C) $\pi/4$ D) $\pi/8$
   E) $8/\pi$ F) 1   G) 1/2   H) 4

   Answer: $4/\pi$ hard
   _____

8. Find the point on the line $y = 2x - 3$ that is nearest to the origin.

   A) $(.5, -2)$ B) $(7.5, -1.5)$ C) $(.875, -1.25)$
   D) $(1, -.5)$ E) $(1.1, -.8)$ F) $(1.2, -.6)$
   G) $(1.25, -.5)$ H) $(1.5, 0)$

   Answer: $(1.2, -.6)$ medium
   _____

9. Find the shortest distance from the point $(1, 4)$ to a point on the parabola $y^2 = 2x$.

A) 1   B) $\sqrt{2}$ C) $\sqrt{3}$ D) 2

E) $\sqrt{5}$ F) $\sqrt{6}$ G) $\sqrt{7}$ H) $2\sqrt{2}$

Answer: $\sqrt{5}$ hard

10. Find the area of the largest rectangle that can be inscribed in the ellipse $\frac{x^2}{a^2} + \frac{y^2}{b^2} = 1$.

A) $ab/2$ B) ab C) $3ab/2$ D) $2ab$

E) $5ab/2$ F) $3ab$ G) $7ab/2$ H) $4ab$

Answer: $2ab$ hard

11. Acme Company's daily profit is given by the following equation: $P(x) = -0.03x^3 + 36x + 500$, $x \geq 0$, where x is the number of units sold each day and $P(x)$ is the daily profit in dollars. How many units would have to be sold to maximize Acme's daily profit?

A) 23   B) 22   C) 21   D) 20

E) 19   F) 18   G) 17   H) 16

Answer: 20 medium

12. A cardboard box of 32 in^3 volume with a square base and open top is to be constructed. Find the minimum area of cardboard needed neglecting waste.

A) 54 in^2 B) 48 in^2 C) 46 in^2 D) 42 in^2

E) 40 in^2 F) 36 in^2 G) 32 in^2 H) 28 in^2

Answer: 48 in^2 medium

13. Farmer Brown wants to fence in a rectangular plot in a large field, using a rock wall which is already there as the north boundary. The fencing for the east and west sides of the plot will cost $3 a yard, but she needs to use special fencing which will cost $5 a yard on the south side of the plot. If the area of the plot is to be 600 square yards, find the dimensions for the plot which will minimize the cost of the fencing. Dimensions below are listed east by south.

A) $10\sqrt{5}$ by $12\sqrt{5}$ yards

B) $12\sqrt{5}$ by $10\sqrt{5}$ yards

C) 10 by 60 yards

D) 60 by 10 yards

E) $8\sqrt{5}$ by $15\sqrt{5}$ yards

F) $15\sqrt{5}$ by $8\sqrt{5}$ yards

G) 15 by 40 yards

H) 40 by 15 yards

Answer: $10\sqrt{5}$ by $12\sqrt{5}$ yards hard

---

14. A closed top container is to be constructed in the shape of a right cylinder. If the surface area is fixed find the ratio of the height to the radius which will maximize the volume.

A) 2:1   B) 3:1   C) 3:2   D) 5:3

E) 1:2   F) 1:3   G) 2:3   H) 3:5

Answer: 2:1 medium

---

15. A gas pipeline is to be constructed from a storage tank, which is right on a road, to a house which is 600 feet down the road and 300 feet back from the road. Pipe laid along the road costs $8.00 per foot, while pipe laid off the road costs $10.00 per foot. What is the minimum cost for which this pipeline can be built? (Assume the pipeline path is piecewise linear, with at most two pieces.)

A) $5000   B) $5500   C) $5900   D) $6100

E) $6600   F) $7000   G) $7100   H) $7500

Answer: $6600 hard

---

16. A plastic right cylinder with closed ends is to hold V cubic feet. If there is no waste in construction, find the ratio between the height and diameter that results in the minimum use of materials.

A) $1:\pi$ B) $1:\pi/2$ C) $1:2\pi$ D) $1:2$

E) $2:\pi$ F) $2:\pi/3$ G) $1:\pi/3$ H) $1:3$

Answer: $1:\pi$ medium

---

17. A right circular cone is inscribed in a hemisphere of radius 2 with the apex of the cone and the center of the base of the hemisphere being the same point. Find the dimensions of one such cone with maximum volume. [NOTE: The volume of a cone is given by $V = \frac{1}{3}\pi r^2 h$.] Dimensions are listed as height by radius.

A) $\frac{2\sqrt{3}}{3}$ by $\frac{2\sqrt{6}}{3}$ B) $\frac{2\sqrt{6}}{3}$ by $\frac{2\sqrt{3}}{3}$ C) $\sqrt{3}$ by $\sqrt{6}$

D) $\frac{\sqrt{3}}{3}$ by $\frac{\sqrt{6}}{3}$ E) $\frac{\sqrt{6}}{3}$ by $\frac{\sqrt{3}}{3}$ F) $\sqrt{6}$ by $\sqrt{3}$

G) $\frac{\sqrt{3}}{2}$ by $\frac{\sqrt{6}}{2}$ H) $\frac{\sqrt{6}}{2}$ by $\frac{\sqrt{3}}{2}$

Answer: $\frac{2\sqrt{3}}{3}$ by $\frac{2\sqrt{6}}{3}$ hard

18. A square is to be cut from each corner of a piece of paper which is 8 cm by 10 cm and the sides are to be folded up to create an open box. What should the side of the square be for maximum volume? (State your answer correct to two decimal places.)

A) 1.35 cm   B) 1.39 cm   C) 1.41 cm   D) 1.47 cm

E) 1.52 cm   F) 1.55 cm   G) 1.60 cm   H) 1.62 cm

Answer: 1.47 cm medium

Calculus, 2nd Edition
by James Stewart
Chapter 3, Section 9
Applications to Economics

1. A company has cost function $C(x) = 1000 + 10x + x^2$. Find the average cost of producing 100 units.

   A) 150   B) 200   C) 210   D) 120

   E) 250   F) 90   G) 180   H) 100

   Answer: 120 easy

2. A company has cost function $C(x) = 1000 + 10x + x^2$. Find the marginal cost of producing 100 units.

   A) 210   B) 120   C) 90   D) 150

   E) 250   F) 200   G) 100   H) 180

   Answer: 210 medium

3. A company has cost function $C(x) = 2000 + 50x + x^2$ and demand function $p(x) = 100$. How many units should it make to maximize its profit?

   A) 15   B) 25   C) 8   D) 5

   E) 30   F) 10   G) 20   H) 35

   Answer: 25 medium

4. A company has cost function $C(x) = 2000 + 50x + x^2$ and demand function $p(x) = 200$. Find the maximum profit the company can make.

   A) 3625   B) 3875   C) 4025   D) 4275

   E) 4500   F) 3945   G) 4225   H) 3600

   Answer: 3625 hard

5. A school band decides to raise money by means of a car wash. Labor and material are donated, so pure profit is realized from sale of the tickets. They know that if they charge $5 per ticket, they will sell 1000, while if they charge $4 per ticket they will sell 1500. Assuming the demand function is linear, what should they charge per ticket to maximize their profit?

    A) $3.75   B) $2.25   C) $2.75   D) $2.50

    E) $3.00   F) $4.00   G) $3.50   H) $3.25

    Answer: $3.50 hard

6. The cost of operating a bus from Azusa to Yreka is $200 + 8x$, where x is the number of passengers. If a ticket is $40, there will be 10 passengers, while if a ticket is $35, there will be 15 passengers. Assuming the demand function is linear, what should the price of a ticket be to maximize profit?

    A) $37   B) $27   C) $29   D) $31

    E) $33   F) $39   G) $35   H) $41

    Answer: $29 hard

7. The cost of operating a bus between Moose Jaw and Saskatoon is $100 + 5x$, where x is the number of passengers. If a ticket is $20, there will be 10 passengers, while if a ticket is $15, there will be 20 passengers. Assuming the demand function is linear, what is the maximum profit possible?

    A) 50   B) 120   C) 100   D) 90

    E) 60   F) 80   G) 110   H) 70

    Answer: 100 hard

8. Suppose that the cost function for an article is given by $C(x) = 0.004x^3 - 0.02x^2 + 6x + 1000$. Assume the cost unit is dollars and find the minimum marginal cost.

    A) $5.87   B) $5.93   C) $5.97   D) $6.01

    E) $6.09   F) $6.14   G) $6.19   H) $6.25

    Answer: $5.97 medium

9. If total cost, c, is related to sales, x, by $c = 0.6x^2 - 179x + 100$, find the amount of sales which will lead to maximum profit.

A) 210   B) 120   C) 90   D) 150

E) 250   F) 200   G) 100   H) 180

Answer: 150 medium

10. If the cost of manufacturing x units per day of a certain commodity is $c = 600 + 0.04x + 0.002x^2$ and if each unit sells for $\$10.00$, what daily production will maximize the profit?

A) 2420 units   B) 2430 units   C) 2440 units

D) 2450 units   E) 2460 units   F) 2470 units

G) 2480 units   H) 2490 units

Answer: 2490 units medium

11. Suppose that $C(x)$ is the number of dollars in the total cost of producing x tables $(x > 6)$ and $C(x) = 25 + 4x + 18x^{-1}$. Find the marginal cost when $x = 10$.

A) $\$3.72$ B) $\$3.75$ C) $\$3.79$ D) $\$3.82$

E) $\$3.89$ F) $\$3.93$ G) $\$3.97$ H) $\$4.00$

Answer: $\$3.82$ medium

Calculus, 2nd Edition
by James Stewart
Chapter 3, Section 10
Antiderivatives

1. Find the most general antiderivative of the function $f(x) = x^2$.

   A) $3x^3 + C$ B) $2x + C$ C) $x/2 + C$ D) $2x^2 + C$

   E) $x^3/3 + C$ F) $3x + C$ G) $x/3 + C$ H) $3x^2 + C$

   Answer: $x^3/3 + C$ (easy)

2. Find the most general antiderivative of $\sec^2 x + \cos x$.

   A) $\frac{1}{3}\sec^3 x + \sin x + C$ B) $\tan^2 x - \sin x + C$ C) $\tan x + \sin x + C$

   D) $\tan x - \sin x + C$ E) $\tan^2 x + \sin x + C$ F) $\frac{1}{3}\sec^3 x + \frac{1}{2}\cos^2 x + C$

   G) $\frac{1}{3}\sec^3 x - \sin x + C$ H) $\frac{1}{3}\sec^3 x - \frac{1}{2}\cos^2 x + C$

   Answer: $\tan x + \sin x + C$ (medium)

3. Given $f'(x) = \sqrt{x}$ and $f(0) = 0$, find $f(1)$.

   A) $1/3$ B) 2 C) $3/2$ D) -2

   E) $2/3$ F) $1/2$ G) $-1/2$ H) 3

   Answer: $2/3$ medium

4. Given $f''(x) = 1$, $f'(0) = 7$, and $f(0) = 3$, find $f(1)$.

   A) $10/3$ B) $17/3$ C) $7/2$ D) $3/10$

   E) $17/2$ F) $13/3$ G) $23/3$ H) $21/2$

   Answer: $21/2$ medium

5. A cyclist traveling at 40 ft/s decelerates at a constant 4 ft/s^2. How many feet does she travel before coming to a complete stop?

   A) 180   B) 200   C) 240   D) 160

   E) 480   F) 400   G) 360   H) 250

   Answer: 200 medium

6. A custard pie is thrown vertically up from the ground with velocity 48 ft/s. Find the greatest distance above the ground that it rises.

   A) 64   B) 72   C) 36   D) 128

   E) 96   F) 48   G) 112   H) 32

   Answer: 36 medium
   _____

7. A chocolate cream pie is thrown vertically up from the ground with velocity 72 ft/s. Find the amount of time in seconds until it hits the ground.

   A) 6.5   B) 4.5   C) 6   D) 7.5

   E) 5.5   F) 7   G) 5   H) 4

   Answer: 4.5 medium
   _____

8. If $f''(x) = -4 \sin 2x$, $f(0) = 0$, $f'(0) = 2$, find the value of $f(\pi/4)$.

   A) $1/\sqrt{2}$ B) $\sqrt{2}$ C) $-1/\sqrt{2}$ D) 0

   E) 1   F) -1 G) $-\sqrt{3}/2$ H) $-\sqrt{2}$

   Answer: 1 medium
   _____

9. If $f'(x) = 3x^2 + 2$ and $f(0) = 1$, find the value of $f(1)$.

   A) 1   B) 2   C) 3   D) 4

   E) 5   F) 6   G) 7   H) 8

   Answer: 4 medium
   _____

10. If $f''(x) = 12x^2 - 2$, $f(0) = 2$, and $f'(0) = 3$, find $f(1)$.

    A) 1   B) 2   C) 3   D) 4

    E) 5   F) 6   G) 7   H) 8

    Answer: 5 medium
    _____

11. Find $f(x)$ if $f'(x) = 3x - x^2$ and $f(1) = 4$.

A) $f(x) = \frac{3}{2}x^2 - \frac{1}{3}x^3 + \frac{17}{6}$

B) $f(x) = \frac{3}{2}x^2 - \frac{1}{3}x^3 + \frac{13}{6}$

C) $f(x) = \frac{3}{2}x^2 - \frac{1}{3}x^3 + \frac{17}{8}$

D) $f(x) = \frac{3}{2}x^2 - \frac{1}{3}x^3 + \frac{13}{8}$

E) $f(x) = 3x^2 - x^3 + \frac{17}{6}$

F) $f(x) = 3x^2 - x^3 + \frac{13}{6}$

G) $f(x) = 3x^2 - x^3 + \frac{17}{8}$

H) $f(x) = 3x^2 - x^3 + \frac{13}{8}$

Answer: $f(x) = \frac{3}{2}x^2 - \frac{1}{3}x^3 + \frac{17}{6}$ medium

12. An astronaut stands on a platform 3 meters above the moon's surface and throws a rock directly upward with an initial velocity of 32 m/s. Given that the acceleration due to gravity on the moon's surface is 1.6 m/s^2 how high, above the surface of the moon, will the rock travel?

A) 317 m   B) 320 m   C) 323 m   D) 326 m

E) 329 m   F) 331 m   G) 334 m   H) 337 m

Answer: 323 m hard

13. Find the position function $s(t)$ given acceleration $a(t) = 3t$ if $v(2) = 0$ and $s(2) = 1$.

A) $s(t) = (t^3/2) - 6t + 9$

B) $s(t) = (t^3/3) - 6t + 9$

C) $s(t) = (t^3/2) - 3t + 9$

D) $s(t) = (t^3/3) - 3t + 9$

E) $s(t) = (t^3/2) - 6t + 1$

F) $s(t) = (t^3/3) - 6t + 1$

G) $s(t) = (t^3/2) - 3t + 1$

H) $s(t) = (t^3/3) - 3t + 1$

Answer: $s(t) = (t^3/2) - 6t + 9$ medium

Calculus, 2nd Edition
by James Stewart
Chapter 4, Section 1
Sigma Notation

1. Find the value of the sum $\sum_{i=0}^{4} i$.

   A) 9   B) 12   C) 6   D) 5

   E) 7   F) 11   G) 8   H) 10

   Answer: 10 easy

2. If we write $1 + \frac{1}{2} + \frac{1}{3} + \frac{1}{4}$ in the sigma notation $\sum_{i=1}^{4}$?, what do we put in place of the question mark?

   A) $i/2$ B) i^{-1} C) $i-1$ D) $2i$

   E) $i+1$ F) i^2 G) i^{-2} H) $1/(1+i)$

   Answer: i^{-1} easy

3. Find the value of the summation $\sum_{i=1}^{5} i^2$.

   A) 55   B) 85   C) 65   D) 25

   E) 95   F) 45   G) 35   H) 75

   Answer: 55 medium

4. Find the value of the summation $\sum_{i=1}^{20} \left(\frac{1}{i} - \frac{1}{1+i}\right)$.

   A) 21/19   B) 20/19   C) 18/19   D) 20/21

   E) 19/18   F) 19/21   G) 21/20   H) 19/20

   Answer: 20/21 medium

5. Find a formula for the sum $\sum_{i=1}^{n} (2i-1)$.

   A) $n^2/2$ B) n^2 C) $2/n^3$ D) $2n^2$

   E) n^3 F) $n^3/2$ G) $2n^3$ H) $2/n^2$

   Answer: n^2 medium

6. Find the value of the sum $\sum_{i=1}^{6} 2^i$.

A) 64   B) 127   C) 128   D) 63

E) 255   F) 191   G) 256   H) 192

Answer: 127 medium

---

7. Find the value of $\sum_{i=1}^{10} (i+1)^2 - \sum_{i=1}^{10} i^2$.

A) 111   B) 144   C) 133   D) 132

E) 145   F) 120   G) 121   H) 110

Answer: 120 medium

---

8. Find the value of the summation $\sum_{k=0}^{8} \cos k\pi$.

A) 0   B) 1/2   C) 1   D) 3/2

E) $\pi/2$ F) $3\pi/2$ G) π H) $2\pi/3$

Answer: 1 medium

---

9. Find the value of the limit $\lim_{n\to\infty} \sum_{i=1}^{n} \frac{1}{n}\left[\left(\frac{i}{n}\right)^3 + 1\right]$.

A) 0   B) 1/4   C) 1/2   D) 1/3

E) 3/4   F) 2/3   G) 1   H) 5/4

Answer: 5/4 medium

---

10. Find the value of the limit $\lim_{n\to\infty} \sum_{i=1}^{n} \frac{3}{n}\left[\left(1+\frac{3i}{n}\right)^3 - 2\left(1+\frac{3i}{n}\right)\right]$.

A) 95/2   B) 191/4   C) 48   D) 193/4

E) 97/2   F) 195/4   G) 49   H) 197/4

Answer: 195/4 hard

Calculus, 2nd Edition
by James Stewart
Chapter 4, Section 2
Area

1. Suppose we wish to approximate the area under the curve $y = x^2$ between $x = 1$ and $x = 3$ using a partition P consisting of 10 equal length subintervals. What is the norm $\|P\|$ of the partition?

    A) 1/2   B) 1/10   C) 1/5   D) 5

    E) 1/20   F) 10   G) 2   H) 1

    Answer: 1/5   easy

2. Suppose we wish to approximate the area under the curve $y = x^2$ between $x = 1$ and $x = 3$ using the partition $P = \{1, 1.5, 1.75, 2, 3\}$. What is the norm $\|P\|$ of the partition?

    A) 3/4   B) 4   C) 3   D) 1/4

    E) 1/8   F) 1   G) 1/2   H) 2

    Answer: 1   medium

3. Suppose we wish to approximate the area under the curve $y = x$ between $x = 1$ and $x = 2$ using a partition consisting of 2 equal-length subintervals. What is the smallest value the approximation could be?

    A) 5/16   B) 0   C) 1/4   D) 3/8

    E) 5/4   F) 7/16   G) 1/2   H) 3/16

    Answer: 5/4   medium

4. Suppose we wish to approximate the area under the curve $y = 2x$ between $x = 3$ and $x = 5$ using a partition consisting of 2 equal-length subintervals. What is the largest value the approximation could be?

    A) 18   B) 10   C) 16   D) 20

    E) 8   F) 14   G) 12   H) 22

    Answer: 18   medium

5. Finding the area under the parabola $y = x^2$ from 0 to 2 can be done by using the limit of a sum

of the form $\lim_{n \to \infty} \frac{c}{n^3} \sum_{i=1}^{n} i^2$. What is the value of the number c?

A) 2   B) 1   C) 4   D) 8

E) 1/6   F) 1/4   G) 1/3   H) 1/2

Answer: 8   hard

6. Finding the area under the parabola $y = x^2$ from 0 to 1 can be done by using the limit of a sum

of the form $\lim_{x \to \infty} \frac{c}{n^3} \sum_{i=1}^{n} i^2$. What is the value of the number c?

A) 8   B) 1/6   C) 1   D) 1/2

E) 4   F) 1/3   G) 2   H) 1/4

Answer: 1   hard

Calculus, 2nd Edition
by James Stewart
Chapter 4, Section 3
The Definite Integral

1. Let $f(x) = x$ on the interval $[1, 2]$. Let the interval be partitioned as follows: $P = \{1, 1.5, 2\}$.

    Find the value of the Riemann sum $\sum_{i=1}^{n} f(x_i^*)\Delta x_i$ if each x_i^* is the left endpoint of its subinterval.

    A) 7/4   B) 5/4   C) 5/2   D) 7/3

    E) 7/6   F) 7/2   G) 5/6   H) 5/3

    Answer: 5/4 easy

2. Let $f(x) = x$ on the interval $[0, 2]$. Let the interval be partitioned as follows: $P = \{0, 1, 2\}$.

    Find the value of the Riemann sum $\sum_{i=1}^{n} f(x_i^*)\Delta x_i$ if each x_i^* is the right endpoint of its interval.

    A) 1/2   B) 5/2   C) 2   D) 7/2

    E) 9/2   F) 4   G) 3/2   H) 3

    Answer: 3 easy

3. Let $f(x) = x$ on the interval $[0, 1]$. Let the interval be partitioned as follows: $P = \{0, 0.5, 1\}$.

    Find the value of the Riemann sum $\sum_{i=1}^{n} f(x_i^*)\Delta x_i$ if each x_i^* is the midpoint of its subinterval.

    A) 3/4   B) 5/8   C) 1   D) 1/2

    E) 1/4   F) 7/8   G) 3/8   H) 1/8

    Answer: 1/2 easy

4. Let $f(x) = x^2$ on the interval $[0, 2]$. Let the interval be partitioned as follows: $P = \{0, 1.5, 2\}$.

    Find the value of the Riemann sum $\sum_{i=1}^{n} f(x_i^*)\Delta x_i$ if each x_i^* is the left endpoint of its subinterval.

    A) 1/2   B) 5/16   C) 7/16   D) 5/8

    E) 3/8   F) 11/16   G) 7/8   H) 9/8

    Answer: 9/8 medium

5.   Let $f(x) = x^2$ on the interval $[0, 2]$. Let the interval be partitioned as follows: $P = \{0, 1, 2\}$.

Find the value of the Riemann sum $\displaystyle\sum_{i=1}^{n} f(x_i^*)\Delta x_i$ if each x_i^* is the midpoint of its subinterval.

A) 5/6   B) 7/6   C) 7/3   D) 7/4

E) 5/2   F) 5/3   G) 5/4   H) 7/2

Answer: 5/2 medium

---

6.   Find the Riemann sum for $f(x) = 16 - x^2$ on the interval $[0, 4]$ if the partition points are $\{0, 1, 2, 3, 4\}$ and right end points are used.

A) 20   B) 22   C) 24   D) 26

E) 28   F) 30   G) 32   H) 34

Answer: 34 medium

---

7.   Use the Midpoint Rule with $n = 5$ to approximate $\displaystyle\int_1^2 \frac{1}{x}\, dx$.

A) 0.6909   B) 0.6913   C) 0.6919   D) 0.6925

E) 0.6928   F) 0.6932   G) 0.6937   H) 0.6945

Answer: 0.6919 medium

---

8.   Use the Midpoint Rule with $n = 4$ to approximate $\displaystyle\int_0^{\pi/4} \tan x\, dx$.

A) 0.2914   B) 0.3160   C) 0.3289   D) 0.3317

E) 0.3450   F) 0.3601   G) 0.3764   H) 0.3844

Answer: 0.3450 medium

Calculus, 2nd Edition
by James Stewart
Chapter 4, Section 4
Properties of the Definite Integral

1. If $\int_0^1 f(x)\,dx = 2$ and $\int_1^2 f(x)\,dx = 1$, find the value of $\int_0^2 f(x)\,dx$.

    A) 3   B) 5   C) 1   D) 6

    E) 2   F) 0   G) 4   H) cannot be determined

    Answer: 3 easy

2. If $\int_0^1 f(x)\,dx = 7$ and $\int_0^3 f(x)\,dx = 6$, find the value of $\int_1^3 f(x)\,dx$.

    A) 1   B) 3   C) 0   D) −3

    E) −1   F) −2   G) 2   H) cannot be determined

    Answer: −1 easy

3. If $\int_0^1 f(x)\,dx = 2$ and $\int_0^3 f(x)\,dx = 3$, find the value of $\int_0^2 f(x)\,dx$.

    A) −2   B) 0   C) 1   D) −1

    E) −3   F) 2   G) 3   H) cannot be determined

    Answer: cannot be determined (medium)

4. If $\int_0^4 f(x)\,dx = 3$ and $\int_0^4 g(x)\,dx = 1$, find the value of $\int_0^4 \left(f(x) + g(x)\right)\,dx$.

    A) 3   B) 10   C) 6   D) 4

    E) 2   F) 8   G) 1   H) cannot be determined

    Answer: 4 easy

5. If $\int_0^4 f(x)\,dx = 5$ and $\int_0^4 g(x)\,dx = 2$, find the value of $\int_0^4 f(x)g(x)\,dx$.

    A) 10   B) 6   C) 1   D) 8

    E) 4   F) 3   G) 2   H) cannot be determined

    Answer: cannot be determined (medium)

6. If $\int_0^3 f(x)\,dx = 4$, $\int_3^6 f(x)\,dx = 4$, and $\int_2^6 f(x)\,dx = 5$, find the value of $\int_0^2 f(x)\,dx$.

A) -3 B) 3 C) 2 D) -1

E) 0 F) 1 G) -2 H) cannot be determined

Answer: 3   medium

7. If $\int_0^3 f(x)\,dx = 12$ and $\int_0^6 f(x)\,dx = 42$ find the value of $\int_3^6 \left(2f(x) - 3\right)\,dx$.

A) 50 B) 51 C) 52 D) 56

E) 53 F) 54 G) 55 H) cannot be determined

Answer: 51   medium

Calculus, 2nd Edition
by James Stewart
Chapter 4, Section 5
The Fundamental Theorem of Calculus

1. Let $f(x) = \int_0^x t^3 \, dt$. Find the value of $f'(2)$.

    A) -24 B) -12 C) 8   D) 24

    E) 4   F) -4 G) 12   H) -8

    Answer: 8   easy

2. Let $f(x) = \int_x^{10} t^3 \, dt$. Find the value of $f'(2)$.

    A) 8   B) 4   C) -24 D) 12

    E) -4 F) -8 G) 24   H) -12

    Answer: -8 easy

3. Let $f(x) = \int_0^{x^2} t^2 \, dt$. Find the value of $f'(1)$.

    A) 4   B) 6   C) 8   D) 0

    E) 2   F) 1   G) 3   H) 5

    Answer: 2   medium

4. Let $f(x) = \int_x^{x^2} t^2 \, dt$. Find the value of $f'(1)$.

    A) 3   B) 8   C) 4   D) 0

    E) 5   F) 2   G) 6   H) 1

    Answer: 1   medium

5. Let $f(x) = \int_{x^3}^{10} (t^2 + 1)^{60} \, dt$. Find the value of $f'(0)$.

    A) 5   B) 1   C) 2   D) 0

    E) 3   F) 6   G) 7   H) 4

    Answer: 0   hard

6. Using the fundamental theorem of calculus, find the value of $\lim\limits_{\|P\|\to 0} \sum\limits_{i=1}^{n} \sqrt{x_i^*}\, \Delta x_i$ on the interval $[1, 4]$.

A) 7/3   B) 10/3   C) 3   D) 13/3

E) 14/3   F) 8/3   G) 4   H) 11/3

Answer: 14/3   medium
_____

7. Find the value of the integral $\int_4^6 x\, dx$.

A) 16   B) 18   C) 15   D) 20

E) 4   F) 8   G) 10   H) 12

Answer: 10   easy
_____

8. Find the value of the integral $\int_0^1 (x+1)^2\, dx$.

A) 7/3   B) 0   C) 1   D) 5/3

E) 2   F) 1/3   G) 4/3   H) 2/3

Answer: 7/3   medium
_____

9. Find the value of the integral $\int_1^2 \frac{1}{x^2}\, dx$.

A) 2/3   B) 1/2   C) −1/2   D) −1/3

E) −2/3   F) 1   G) 1/3   H) −1

Answer: 1/2   medium
_____

10. Find the value of the integral $\int_{-1}^8 \sqrt[3]{x}\, dx$.

A) 49/4   B) 41/4   C) 45/4   D) 35/4

E) 39/4   F) 37/4   G) 47/4   H) 43/4

Answer: 45/4   medium
_____

11. Find the value of the integral $\int_{-3}^3 (|x| + |x+1|)\, dx$.

A) 19   B) 16   C) 11   D) 0

E) 9   F) 17   G) 13   H) 12

Answer: 19   hard
_____

12. If $F(x) = \int_0^{\sqrt{x}} \sin(t^2)\, dt$, find $F'(\pi/4)$.

   A) $1/\sqrt{2}$ B) π C) $\sqrt{\pi/2}$ D) $\sqrt{2\pi}$

   E) $1/\sqrt{2\pi}$ F) $\pi/2$ G) $\pi^2/16$ H) $\sin(\pi^2/16)$

   Answer: $1/\sqrt{2\pi}$ medium

13. Find the value of the integral $\int_{-1}^{1} (1+x)^2\, dx$.

   A) 1/3   B) $-8/3$ C) 3/8   D) 0

   E) 2/3   F) 8/3   G) 4/3   H) 3

   Answer: 8/3   medium

14. If $F(x) = \int_x^{x^2} \ln \sqrt{t}\, dt$, find the value of $F'(1/4)$.

   A) 0   B) 1   C) 3/2   D) 2

   E) 5/2   F) 3   G) 7/2   H) 4

   Answer: 0   hard; requires Chapter 6

15. Find the value of the integral $\int_{-2}^{0} |x+1|\, dx$.

   A) 1/2   B) 1/4   C) 3/4   D) 3/2

   E) 5/4   F) 5/2   G) 1   H) 0

   Answer: 1   medium

16. If $F(x) = \int_0^{x^2} \sqrt{1+8t^3}\, dt$, find the value of $F'(1)$.

   A) 1   B) 2   C) 3   D) 4

   E) 5   F) 6   G) 7   H) 8

   Answer: 6   medium

17. If $\int_3^{b} 3x^2\, dx = 37$, find the value of b.

   A) 4   B) 5   C) 6   D) 7

   E) 8   F) 9   G) 10   H) 11

   Answer: 4   medium

18. If $F(x) = \displaystyle\int_2^{1/x} \sin^4 t \, dt$, find $F'(2/\pi)$.

    A) $-\pi^2/4$ B) $\pi^2/4$ C) $2 - \pi^2/4$ D) 1

    E) 0   F) $\sin^4 2$ G) $1/2$ H) does not exist

    Answer: $-\pi^2/4$ hard

19. If $F(x) = \displaystyle\int_2^{3x} e^{t^4} \, dt$, find the value of $F'(0)$.

    A) 1   B) 2   C) 3   D) 4

    E) 6   F) 8   G) e^8 H) e^{16}

    Answer: 3   medium; requires Chapter 6

20. Find the value of the integral $\displaystyle\int_0^1 (1 + \sqrt{x})^2 \, dx$.

    A) $17/6$ B) 3   C) $19/6$ D) $10/3$

    E) $7/2$ F) $11/3$ G) $23/6$ H) 4

    Answer: $17/6$ medium

21. If $F(x) = \displaystyle\int_0^{\sqrt{x}} \frac{1}{\sqrt{1 + t^4}} \, dt$, find the value of $F'(1)$.

    A) $1/2$ B) $\sqrt{2}/2$ C) $\sqrt{2}/4$ D) $1/4$

    E) $\sqrt{2}/8$ F) $1/8$ G) $\sqrt{2}/16$ H) $1/16$

    Answer: $\sqrt{2}/4$ medium

22. If $F(x) = \displaystyle\int_1^{\sqrt{x}} \frac{e^t}{t} \, dt$, find the value of $F'(1)$.

    A) 0   B) $1/3$ C) $1/2$ D) 1

    E) $e/3$ F) $e/2$ G) e H) e^2

    Answer: $e/2$ medium; requires Chapter 6

23. If $F(x) = \displaystyle\int_0^{\sqrt{x}} \sqrt{t^4 + 20} \, dt$, find the value of $F'(4)$.

    A) $1/2$ B) 1   C) $3/2$ D) 2

    E) $5/2$ F) 3   G) $7/2$ H) 4

    Answer: $3/2$ medium

24. Find the value of the integral $\int_1^2 \frac{x^2-1}{x}\,dx$.

A) 1/2

B) 1

C) 3/2

D) 2

E) $(1/2)\ln 2$

F) $1+\ln 2$

G) $(3/2)-\ln 2$

H) $2+\ln 2$

Answer: $(3/2)-\ln 2$ (medium; requires Chapter 6)

---

25. The velocity of a particle moving along a line is $2t$ meters per second. Find the distance traveled in meters during the time interval $1 \le t \le 3$.

A) 9

B) 5

C) 2

D) 8

E) 4

F) 3

G) 6

H) 7

Answer: 8   easy

---

26. The velocity of a particle moving along a line is $t^3 - t$ meters per second. Find the distance traveled in meters during the time interval $0 \le t \le 2$.

A) 7/4

B) 4/3

C) 2/3

D) 3/4

E) 3/2

F) 2/5

G) 5/2

H) 9/4

Answer: 5/2   medium

---

27. The acceleration of a particle moving along a line is $\sqrt{t}$ meters per second. It starts from a resting position at $t = 0$. Find the distance traveled in meters during the time interval $0 \le t \le 1$.

A) 3/4

B) 7/4

C) 3/2

D) 4/3

E) 2/5

F) 5/2

G) 2/3

H) 4/15

Answer: 4/15   medium

28. Let $y = \displaystyle\int_1^{3x} \frac{dt}{t^2 + t + 1}$. Find $\dfrac{d^2 y}{dx^2}$.

A) $\dfrac{(18x + 3)}{(9x^2 + 3x + 1)^2}$

B) $\dfrac{3(18x + 3)}{(9x^2 + 3x + 1)^2}$

C) $\dfrac{-3(18x + 3)}{(9x^2 + 3x + 1)^2}$

D) $\dfrac{3(18x + 3)}{(9x^2 + 3x + 1)^3}$

E) $\dfrac{-3(18x + 3)}{(9x^2 + 3x + 1)^3}$

F) $\dfrac{(18x + 3)}{(9x^2 + 3x + 1)^3}$

G) $\dfrac{(18x + 3)^2}{(9x^2 + 3x + 1)^3}$

H) $\dfrac{-3(18x + 3)^2}{(9x^2 + 3x + 1)^2}$

Answer: $\dfrac{-3(18x + 3)}{(9x^2 + 3x + 1)^2}$ medium

29. Let $f(x) = \displaystyle\int_0^x \frac{t^2 - 4}{1 + \cos^2 t}\, dt$. At what value of x does the local maximum of $f(x)$ occur?

A) -4 B) -3 C) -2 D) -1

E) 0 F) 1 G) 2 H) 3

Answer: -2 hard

30. Evaluate $\dfrac{d}{dt} \displaystyle\int_{t^2}^2 \sqrt{x + 1}\, dx$.

A) $\sqrt{t^2 + 1}$

B) $-\sqrt{t^2 + 1}$

C) $2t$

D) $\sqrt{t + 1}$

E) $2t\left(\sqrt{t^2 + 1}\right)$

F) $2t\left(-\sqrt{t^2 + 1}\right)$

G) $2t\left(\sqrt{t + 1}\right)$

H) $2t\left(-\sqrt{t + 1}\right)$

Answer: $2t\left(-\sqrt{t^2 + 1}\right)$ medium

31. Evaluate $\displaystyle\int_{-3}^4 \bigl||x| - 4\bigr|\, dx$.

A) $29/2$ B) 15 C) $31/2$ D) 16

E) $35/2$ F) $33/2$ G) 17 H) cannot be determined

Answer: $31/2$ medium

32. Solve for x when $\displaystyle\int_0^1 t^x \, dt = 5$.

A) $-4/5$ B) $-3/5$ C) $-2/5$ D) $-1/5$

E) 0   F) $1/5$ G) $2/5$ H) $3/5$

Answer: $-4/5$ medium

33. If $\displaystyle\int_0^5 kx \, dx = 30$, find k.

A) $12/5$ B) $11/5$ C) 2   D) $9/5$

E) $8/5$ F) $7/5$ G) $6/5$ H) 1

Answer: $12/5$ medium

Calculus, 2nd Edition
by James Stewart
Chapter 4, Section 6
The Substitution Rule

1. Find the value of the integral $\int_0^3 \sqrt{x+1} \, dx$.

    A) 13/3 B) 5 C) 6 D) 16/3

    E) 20/3 F) 14/3 G) 19/3 H) 17/3

    Answer: 14/3 easy

2. Find the value of the integral $\int_0^{\pi/4} \cos 2x \, dx$.

    A) $\pi/14$ B) π C) $\sqrt{3}/2$ D) $\pi/2$

    E) $\sqrt{2}/4$ F) 2π G) $\sqrt{2}/2$ H) 1/2

    Answer: 1/2 easy

3. Find the value of the integral $\int_0^1 (2x)^7 \, dx$.

    A) 64 B) 32 C) 2 D) 128

    E) 256 F) 8 G) 4 H) 16

    Answer: 16 easy

4. Find the value of the integral $\int_0^1 (x^2+1)^5 \, x \, dx$.

    A) 13/2 B) 17/2 C) 17/3 D) 29/4

    E) 27/5 F) 21/4 G) 25/17 H) 28/3

    Answer: 21/4 medium

5. Find the value of the integral $\int_0^1 \frac{x^2}{(x^3+1)^2} \, dx$.

    A) 3/4 B) 2 C) 3/7 D) 7/3

    E) 1/6 F) 3/2 G) 2/3 H) 1

    Answer: 1/6 medium

6. Find the value of the integral $\int_0^{\pi/4} \sin^2 x \cos x \, dx$.

A) $\sqrt{2}/18$ B) $\pi/4$ C) $3\pi/2$ D) $\sqrt{2}/12$

E) $\sqrt{2}/6$ F) $\sqrt{2}/9$ G) $2\pi/3$ H) $\pi/2$

Answer: $\sqrt{2}/12$ medium

7. Find the value of the integral $\int_0^{\pi/4} \sin 2x \sin x \, dx$.

A) $\sqrt{2}/9$ B) $\sqrt{2}/6$ C) $2\pi/3$ D) $\sqrt{2}/12$

E) $\pi/4$ F) $3\pi/2$ G) $\sqrt{2}/18$ H) $\pi/2$

Answer: $\sqrt{2}/6$ hard

8. Find the value of the integral $\int_0^{\pi/3} \sec x \tan x \, (1 + \sec x) \, dx$.

A) 4   B) 5/2   C) 3   D) 11/2

E) 9/2   F) 2   G) 7/2   H) 5

Answer: 5/2   medium

9. Find the value of the integral $\int_1^4 \frac{1}{(1 + \sqrt{x})^2} \frac{1}{\sqrt{x}} \, dx$.

A) 6/5   B) 1/3   C) 2/3   D) 5/2

E) 4/9   F) 3/2   G) 5/6   H) 1/6

Answer: 1/3   hard

10. Find the value of the integral $\int_0^1 \frac{x}{x+1} \, dx$.

A) $1 - \ln 2$ B) $\ln 4$ C) e^2 D) $e - 1$

E) $e^2 - 1$ F) $1/e$ G) $\ln 4 - 1$ H) $\ln 6$

Answer: $1 - \ln 2$ (hard; requires Chapter 6)

11. Find the value of the integral $\int_0^{\pi/2} \cos x \sin(\sin x) \, dx$.

A) $\pi/2$ B) $1 - (\pi/4)$ C) $\sin 1$

D) $1 - \cos 1$ E) $(\pi/2) - \sin 1$ F) $(\pi/4) + \cos 1$

G) $1 + (3\pi/4)$ H) $1 + \tan 1$

Answer: $1 - \cos 1$ (medium)

12. Find the value of the integral $\int_{-1}^{1} \dfrac{\tan x}{1+x^2}\, dx$.

A) 0 B) $\pi/2$ C) π D) 2

E) -2 F) $\ln 2$ G) $1+\ln 2$ H) does not exist

Answer: 0 medium

---

13. Find the value of the integral $\int_{-1}^{1} \dfrac{x}{\sqrt{x^2+1}}\, dx$.

A) $2(\sqrt{2}-1)$ B) $\sqrt{2}-1$ C) 0 D) 1

E) 3 F) $1/\sqrt{2}$ G) $\sqrt{2}$ H) does not exist

Answer: 0 medium

---

14. Find the value of the integral $\int_{0}^{1} \dfrac{x}{\sqrt{x^2+1}}\, dx$.

A) $2(\sqrt{2}-1)$ B) $\sqrt{2}-1$ C) 0 D) 1

E) 3 F) $1/\sqrt{2}$ G) $\sqrt{2}$ H) does not exist

Answer: $\sqrt{2}-1$ medium

---

15. Find the value of the integral $\int_{0}^{\pi/2} \dfrac{\cos x}{1+\sin^2 x}\, dx$.

A) 0 B) $\pi/4$ C) $\pi/2$ D) $3\pi/4$

E) π F) $3\pi/2$ G) 2π H) $5\pi/2$

Answer: $\pi/4$ hard; requires Chapter 6

---

16. Find the value of the integral $\int_{0}^{\pi/4} \sec^2 x \, \sin(\tan x)\, dx$.

A) $\pi/2$ B) $1-\pi/4$ C) $\sin 1$

D) $1-\cos 1$ E) $(\pi/2)-\sin 1$ F) $(\pi/4)+\cos 1$

G) $1+(3\pi/4)$ H) $1+\tan 1$

Answer: $1-\cos 1$ medium

---

17. Find the value of the integral $\int_{-\pi/2}^{\pi/2} \dfrac{x^2 \sin x}{1+x^6}\, dx$.

A) 0 B) $1/2$ C) 1 D) 2

E) $\pi/6$ F) $\pi^2/4$ G) $(\pi^3/3)-1$ H) $\pi/2$

Answer: 0 medium

---

18. Find the value of the integral $\int_{1}^{3} \dfrac{x}{\sqrt{1+3x^2}}\, dx$.

A) $(2/3)(\sqrt{7}-1)$ B) $\sqrt{7}-1$ C) 0

D) $1/3$ E) $2/3$ F) $1/\sqrt{7}$

G) $\sqrt{7}$ H) does not exist

Answer: $(2/3)(\sqrt{7}-1)$ medium

19. Find the value of the integral $\int_{1}^{9} \dfrac{3x}{\sqrt{10-x}}\, dx$.

A) 68 B) 126 C) 0 D) 34

E) 58 F) 27 G) 26 H) does not exist

Answer: 68 hard

20. Find the value of the integral $\int_{1}^{8} \dfrac{4\left(x^{2/3}+14\right)^{3}}{\sqrt[3]{x}}\, dx$.

A) $163001/2$ B) $163053/2$ C) 81500

D) 81527 E) 3976 F) 7952

G) $15905/2$ H) $16783/2$

Answer: $163053/2$ hard

21. Evaluate the integral $\int \dfrac{r}{\left(r^2+b^2\right)^{3/2}}\, dr$ where b is a constant.

A) r^2+b^2+C B) $\left(r^2+b^2\right)^{1/2}+C$ C) $-\left(r^2+b^2\right)+C$

D) $-\left(r^2+b^2\right)^{1/2}+C$ E) $\left(r^2+b^2\right)^{-1/2}+C$ F) $-\left(r^2+b^2\right)^{-1/2}+C$

G) $\left(r^2+b^2\right)^{3/2}+C$ H) $-\left(r^2+b^2\right)^{3/2}+C$

Answer: $-\left(r^2+b^2\right)^{-1/2}+C$ medium

Calculus, 2nd Edition
by James Stewart
Chapter 5, Section 1
Areas between Curves

1.   Find the area of the region bounded by the curves $y = x^2$ and $y = 1$.

   A) 4/3   B) 2/3   C) 11/9   D) 5/3

   E) 8/9   F) 1   G) 14/9   H) 2

   Answer: 4/3 easy
   _____

2.   Find the area of the region bounded by the curves $y = x^2$ and $y = x$.

   A) 1/8   B) 2/3   C) 1/12   D) 1/2

   E) 1/3   F) 1/9   G) 1/6   H) 5/6

   Answer: 1/6 easy
   _____

3.   Find the area of the region bounded by the curves $y = x^2$ and $y = 1 - x^2$.

   A) $\sqrt{2}/2$ B) $\sqrt{3}/6$ C) $\sqrt{3}/2$ D) $2\sqrt{2}/3$

   E) $\sqrt{3}/3$ F) $\sqrt{3}/4$ G) $\sqrt{2}/4$ H) $\sqrt{2}/6$

   Answer: $2\sqrt{2}/3$ medium
   _____

4.   Find the area of the region bounded by the curves $y = x^2$ and $y = x^3$.

   A) 1/4   B) 1/9   C) 1/18   D) 1/15

   E) 1/12   F) 1/6   G) 1/24   H) 1/3

   Answer: 1/12 medium
   _____

5.   Find the area of the region bounded by the curves $y = x^2$ and $x = y^2$.

   A) 1/6   B) 1/4   C) 1/15   D) 1/18

   E) 1/9   F) 1/3   G) 1/12   H) 1/24

   Answer: 1/3 medium
   _____

6. Find the area of the region bounded by the curves $y = x$ and $y = x^3$.

   A) 5/3   B) 2/3   C) 3/2   D) 4/3
   E) 1/3   F) 1/2   G) 1   H) 3/4

   Answer: 1/2 hard
   _____

7. Find the area beneath a single arc of the function $y = \sin x$ (for example, over the interval $[0, \pi]$).

   A) π B) 1/4   C) 1   D) 2π
   E) $\pi/2$ F) $\pi/4$ G) 1/2   H) 2

   Answer: 2 medium
   _____

8. Find the area of the region bounded by the curves $x = 1 - y^4$ and $x = y^3 - y$.

   A) 1   B) 6/5   C) 7/5   D) 8/5
   E) 9/5   F) 2   G) 11/5   H) 12/5

   Answer: 8/5 hard
   _____

9. Find the area of the region bounded by the parabolas $y = 2x - x^2$ and $y = x^2$.

   A) 1/2   B) 2/3   C) 3/4   D) 2/5
   E) 1/3   F) 1/4   G) 1/5   H) 3/5

   Answer: 1/3 medium
   _____

10. Find the area of the region bounded by the parabolas $y = 4x^2$ and $y = x^2 + 3$.

    A) 1/2   B) 1   C) 3/2   D) 2
    E) 5/2   F) 3   G) 7/2   H) 4

    Answer: 4 medium
    _____

11. Find the area of the region bounded by the parabola $x = y^2$ and the line $x - 2y = 3$.

    A) 29/3   B) 32/3   C) 35/3   D) 38/3
    E) 41/3   F) 44/3   G) 47/3   H) 50/3

    Answer: 32/3 medium
    _____

12. Find an expression as a limit of Riemann sums for the area between the curves $y = -x$ and $y = x - x^2$, using equal subintervals and right end points.

A) $\lim\limits_{n\to\infty} \frac{2}{n} \sum\limits_{i=1}^{n} \left[2\left(\frac{2i}{n}\right) - \left(\frac{2i}{n}\right)^2 \right]$

B) $\lim\limits_{n\to\infty} \frac{1}{n} \sum\limits_{i=1}^{n} \left(\frac{i}{n}\right)^2$

C) $\lim\limits_{n\to\infty} \frac{1}{n} \sum\limits_{i=1}^{n} \left[\left(\frac{i}{n}\right)^2 - \left(\frac{i}{n}\right) \right]$

D) $\lim\limits_{n\to\infty} \frac{2}{n} \sum\limits_{i=1}^{n} \left[-1 + \left(\frac{2i}{n}\right)^2 \right]$

E) $\lim\limits_{n\to\infty} \frac{2}{n} \sum\limits_{i=1}^{n} \left[2\left(\frac{2i}{n}\right)^2 - \left(\frac{2i}{n}\right) \right]$

F) $\lim\limits_{n\to\infty} \frac{1}{n} \sum\limits_{i=1}^{n} \left[2\left(\frac{2i}{n}\right) - \left(\frac{2i}{n}\right)^2 \right]$

G) $\lim\limits_{n\to\infty} \frac{1}{n} \sum\limits_{i=1}^{n} \left[\left(-1 + \left(\frac{i}{n}\right)^2\right) - 1 \right]$

H) $\lim\limits_{n\to\infty} \sum\limits_{i=1}^{n} \left[\left(-1 + \left(\frac{2i}{n}\right)^2\right) + 3\left(-1 + \left(\frac{2i}{n}\right)\right) \right]$

Answer: $\lim\limits_{n\to\infty} \frac{2}{n} \sum\limits_{i=1}^{n} \left[2\left(\frac{2i}{n}\right) - \left(\frac{2i}{n}\right)^2 \right]$ (hard)

13. Find the area of the region bounded by the curves $f(y) = x^3 + x^2$ and $g(y) = 2x^2 + 2x$.

Answer: $\frac{37}{12}$ medium

14. Find the area of the region bounded by the curves $f(y) = x^2 - 4x + 5$ and $g(y) = 5 - x$.

Answer: $\frac{9}{2}$ easy

15. Find the area of the region bounded by the curves $f(y) = x^2 - 1$ and $g(y) = -x^2 + x + 2$.

Answer: $\frac{125}{24}$ medium

16. Find the area of the region bounded by the curves $f(y) = 4 - x^2$ and $g(y) = x^2 + 2$.

Answer: $\frac{8}{3}$ easy

17. Find the area of the region bounded by the curves $f(y) = \sqrt{x}$, $g(y) = \frac{5-x}{4}$, and $h(y) = \frac{3x-8}{2}$.

Answer: $\frac{23}{12}$ medium

18. Find the area of the region bounded by the curves $f(y) = x^3 - x^2 - 2x$ and $g(y) = 0$.

Answer: $\frac{37}{12}$ medium

_____

19. Find the area of the region bounded by the curves $f(y) = x^2 + 2x - 1$ and $g(y) = x^3 - 1$.

Answer: $\frac{37}{12}$ medium

_____

20. Find the area of the region bounded by the x-axis and the graph of $y = 6x(x-2)^2$.

Answer: 8 easy

_____

21. Find the area of the region bounded by the curves $x = y^2 - 7$ and $x = y - 1$.

Answer: $\frac{125}{6}$ medium

_____

22. Let R be the region bounded by: $y = x^3$, the tangent to $y = x^3$ at $(1, 1)$, and the x-axis. Find the area of R integrating a) with respect to x, b) with respect to y.

Answer: a) and b) each $\frac{1}{12}$ medium

_____

23. Find the area of the region bounded by the curves $f(y) = x^2 - 2x$ and $g(y) = -2x^2 + 3x + 2$.

Answer: $5\frac{35}{54}$ medium

_____

24. Find the area of the region bounded by the curves $f(y) = x^2$ and $g(y) = x + 1$.

Answer: $\frac{5\sqrt{5}}{6}$ medium

_____

Calculus, 2nd Edition
by James Stewart
Chapter 5, Section 2
Volumes

1. Find the volume of the solid obtained when the region bounded by the x-axis, the y-axis, and the line $y - x = 3$ is rotated about the x-axis.

   A) 3π B) 2π C) 8π D) 10π

   E) 9π F) 12π G) 4π H) 6π

   Answer: 9π easy

2. Find the volume of the solid obtained when the region bounded by the line $y = x$, the line $x = 3$, and the x-axis is rotated about the y-axis.

   A) 36π B) 42π C) 16π D) 18π

   E) 24π F) 30π G) 48π H) 27π

   Answer: 18π medium

3. Suppose the disk method is used to find the volume of the solid obtained when the region bounded by the curve $x = y^2$ and the line $x = 4$ is rotated about the x-axis. What is the area of the largest cross-section?

   A) 6π B) 3π C) 12π D) 10π

   E) 9π F) 8π G) 2π H) 4π

   Answer: 4π easy

4. Find the volume of the solid obtained when the region bounded by the curve $x = y^2$ and the line $x = 4$ is rotated about the x-axis.

   A) 10π B) 4π C) 9π D) 12π

   E) 6π F) 8π G) 3π H) 2π

   Answer: 8π medium

5. Find the volume of the solid obtained when the region bounded by the curves $y = x^2/4$ and $x = y^2/4$ is rotated about the x-axis.

A) 15.6π B) 16.8π C) 18.2π D) 17.8π

E) 15.8π F) 19.2π G) 17.3π H) 18.4π

Answer: 19.2π medium

6. Find the volume of the solid obtained when the region bounded by the curve $y = \sin x$, $0 \le x \le \pi$, and the x-axis is rotated about the x-axis.

A) $\pi^2/2$ B) $\pi^2/3$ C) π D) $\pi/2$

E) $\pi/4$ F) π^2 G) $\pi/3$ H) $\pi^2/4$

Answer: $\pi^2/2$ medium

7. Find the volume of the solid obtained when the region bounded by the curve $y = x^2$ and the line $y = 4$ is rotated about the line $y = 4$.

A) $32\pi/15$ B) $32\pi/5$ C) $128\pi/15$ D) $64\pi/5$

E) $64\pi/15$ F) $512\pi/15$ G) $256\pi/5$ H) $128\pi/5$

Answer: $512\pi/15$ hard

8. The base of a solid S is the parabolic region $\{(x, y) \mid y^2 \le x \le 1\}$. Cross-sections perpendicular to the x-axis are squares. Find the volume of S.

A) 1.5   B) 1.6   C) 1.7   D) 1.8

E) 1.9   F) 2.0   G) 2.1   H) 2.2

Answer: 2.0 medium

9. Find the volume of the solid obtained by rotating the region bounded by the curves $y = \sqrt{x}$ and $y = x^2$ about the line $y = 3$.

A) 1.5π B) 1.6π C) 1.7π D) 1.8π

E) 1.9π F) 2π G) 2.1π H) 2.2π

Answer: 1.7π medium

10. A hole of radius 4 is bored through the center of a sphere of radius 5. Find the volume of the remaining portion of the sphere.

A) 16π B) 20π C) 24π D) 28π

E) 32π F) 36π G) 40π H) 44π

Answer: 36π hard

11. Find the volume of the solid obtained by rotating the region bounded by the curves $y = x$ and $y = x^2$ about the line $y = 2$.

A) $\pi/5$ B) $4\pi/15$ C) $\pi/3$ D) $2\pi/5$

E) $7\pi/15$ F) $8\pi/15$ G) $3\pi/5$ H) $2\pi/3$

Answer: $8\pi/15$ medium

12. The base of S is a circular disc with radius 3. Parallel cross-sections perpendicular to the base are isosceles triangles with height 2 and unequal side in the base. Find the volume of S.

A) 3π B) 4π C) 5π D) 6π

E) 7π F) 8π G) 9π H) 10π

Answer: 9π hard

13. Find the volume of the solid obtained by rotating about the line $y = 1$ the region bounded by $y = \cos x$, $y = 0$, $x = 0$, and $x = \pi/2$.

A) π B) π^2 C) $\pi^2 - (\pi/2)$

D) $\pi^2 - \pi$ E) $\pi^2 - 2\pi$ F) $2\pi - (\pi^2/4)$

G) $2\pi - (\pi^2/2)$ H) $\pi - (\pi^2/4)$

Answer: $2\pi - (\pi^2/4)$ hard

14. The region bounded by the curves $y = x - x^2$ and $y = 0$ is rotated about the x-axis. Find the volume of the resulting solid.

A) $\pi/3$ B) $\pi/4$ C) $\pi/5$ D) $\pi/6$

E) $\pi/12$ F) $\pi/24$ G) $\pi/30$ H) $\pi/36$

Answer: $\pi/30$ medium

15. Find the volume of the solid obtained by rotating the region bounded by the curves $y = 2 - x^2$ and $y = 1$ about the x-axis.

A) $\pi/15$ B) $2 - \pi^2$ C) 15π D) $7/15$

E) $3 - \pi^2$ F) 4π G) $56\pi/15$ H) $128\pi/15$

Answer: $56\pi/15$ medium
_____

16. Find the volume of the solid obtained by rotating about the x-axis the region bounded by $y = x^2$, $y = 0$, and $x = 1$.

A) π B) $\pi/2$ C) $\pi/3$ D) $\pi/4$

E) $\pi/5$ F) $\pi/6$ G) $\pi/7$ H) $\pi/8$

Answer: $\pi/5$ medium
_____

17. A solid has a circular base of radius 1. Parallel cross-sections perpendicular to the base are equilateral triangles. Find the volume of the solid.

A) $\pi/2$ B) $\sqrt{3}/2$ C) $3\pi/2$ D) $3\sqrt{3}/2$

E) $2\pi/3$ F) $4\sqrt{3}/3$ G) $3\pi/4$ H) $2\sqrt{2}/3$

Answer: $4\sqrt{3}/3$ medium
_____

18. Find the volume of the solid obtained by rotating about the x-axis the region bounded by the curves $y = x$ and $y = x^2$.

A) $\pi/30$ B) $2\pi/15$ C) $7\pi/30$ D) $\pi/3$

E) $13\pi/30$ F) $8\pi/15$ G) $19\pi/30$ H) $11\pi/15$

Answer: $2\pi/15$ medium
_____

19. The base of a solid S is a semi-circular disc $\{(x, y) \mid x^2 + y^2 \leq 1, \ x \geq 0\}$. Cross-sections of S perpendicular to the x-axis are squares. Find the volume of S.

A) $2\pi/3$ B) π^2 C) $\pi^2/4$ D) $8/3$

E) $1/3$ F) 1 G) π H) 2

Answer: $8/3$ medium
_____

20. The volume of a sphere of radius r is $\frac{4\pi r^3}{3}$. Verify this formula by revolving the circle $x^2 + y^2 = r^2$ about the x-axis.

Answer: Volume $= 2\int_0^r \pi y^2 \, dx = 2\pi \int_0^r \left(r^2 - x^2\right) dx$

$$= 2\pi\left(r^3 x - \tfrac{1}{3}x^3\right)\Big|_0^r$$

$$= 2\pi\left(r^3 - \tfrac{1}{3}r^3\right) = 2\pi\left(\tfrac{2}{3}r^3\right) = \tfrac{4}{3}\pi r^3 \quad \text{medium}$$

21. Find the volume of the solid generated by revolving about the line $y = 4$ the smaller region bounded by the curve $x^2 = 4y$ and the lines $x = 2$ and $y = 4$.

Answer: $\frac{106}{15}\pi$ easy

22. Consider the region in the xy-plane between $x = 0$ and $x = \pi/2$, bounded by $y = 0$ and $y = \sin x$. Find the volume of the solid generated by revolving this region about the x-axis.

Answer: $\frac{\pi^2}{4}$ medium

23. Find the volume of the solid formed when the region bounded by the curves $y = x^3 + 1$, $x = 1$ and $y = 0$ is rotated about the x-axis.

Answer: $\frac{23}{14}\pi$ easy

24. Find the volume of the solid obtained when the region enclosed by $y = x^2$ and $y = 2 - x^2$ is revolved about the x-axis.

Answer: $\frac{16\pi}{3}$ medium

25. Find the volume of the solid generated by revolving about the line $y = -1$ the region bounded by the graphs of the equations $y = x^2 - 4x + 5$ and $y = 5 - x$.

Answer: $\frac{162\pi}{5}$ medium

26. A curve described by the equation $(x-1)^2 = 32 - 3y$ is rotated about the line $x = 1$ to generate a solid of revolution. Find the volume of this solid for the region bounded by $x = 1$, $y = 1$, and $y = 4$ and to the right of $x = 1$.

Answer: $\dfrac{141\pi}{2}$ medium

27. Find the volume of the solid of revolution obtained by revolving the region bounded by $y = x^2$, the x-axis, and $x = 2$ around the line $y = -1$.

Answer: $\dfrac{176\pi}{15}$ medium

28. Consider the plane region bounded by the curves given by $y = x^2 + 1$ and $y = x + 3$. Find the volume of the solid of revolution generated by revolving the region around the x-axis.

Answer: $\dfrac{117\pi}{5}$ easy

29. Find the volume of one octant of the region common to two right circular cylinders of radius 1 whose axes intersect at right angles as shown below:

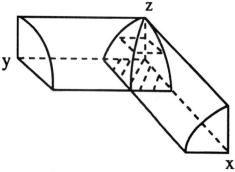

Answer: $\dfrac{2}{3}$ hard

30. Find the volume of the solid generated when the region in the first quadrant bounded by the curve $y = 4x - x^3$ and the x-axis is rotated about the x-axis.

Answer: $\dfrac{1024\pi}{105}$ easy

Calculus, 2nd Edition
by James Stewart
Chapter 5, Section 3
Volumes by Cylindrical Shells

1. Find the volume of the solid obtained when the region bounded by the curve $y = 2x - x^2$ and the x-axis is rotated about the y-axis.

   A) $10\pi/3$ B) $20\pi/3$ C) $8\pi/3$ D) $4\pi/3$

   E) $16\pi/3$ F) $32\pi/3$ G) $5\pi/3$ H) $40\pi/3$

   Answer: $8\pi/3$ medium
   _____

2. Find the volume of the solid obtained when the region bounded by the curve $y = \sin x$, $0 \le x \le \pi$, and the x-axis is rotated about the y-axis.

   A) $\pi^3/2$ B) π^2 C) π^3 D) $4\pi^2$

   E) $2\pi^2$ F) $2\pi^3$ G) $\pi^2/2$ H) $4\pi^3$

   Answer: $2\pi^2$ hard
   _____

3. Find the volume of the solid obtained when the region above the x-axis bounded by the x-axis and the curve $y = x - x^3$ is rotated about the y-axis.

   A) $2\pi/7$ B) $8\pi/35$ C) $4\pi/15$ D) $8\pi/15$

   E) $4\pi/35$ F) $2\pi/5$ G) $4\pi/5$ H) $4\pi/7$

   Answer: $4\pi/15$ medium
   _____

4. Find the volume of the solid obtained when the region bounded by the curves $y = 2x - x^2$ and $y = x^2 - 2x$ is rotated about the y-axis.

   A) $40\pi/3$ B) $8\pi/3$ C) $20\pi/3$ D) $16\pi/3$

   E) $4\pi/3$ F) $10\pi/3$ G) $5\pi/3$ H) $32\pi/3$

   Answer: $16\pi/3$ medium
   _____

5. Find the volume of the solid obtained when the region bounded by the lines $y = x$, $x = 1$, $x = 2$, and the x-axis is rotated about the y-axis.

    A) $16\pi/3$ B) 8π C) $20\pi/3$ D) $22\pi/3$

    E) 6π F) $14\pi/3$ G) 4π H) $10\pi/3$

    Answer: $14\pi/3$ medium
    _____

6. Find the volume of the solid obtained when the region bounded by the curve $y = 5x^3$, $x = 1$, $x = 2$, and the x-axis is rotated about the y-axis.

    A) $127\pi/3$ B) $62\pi/3$ C) 31π D) 15π

    E) 5π F) 127π G) 62π H) $31\pi/3$

    Answer: 62π medium
    _____

7. Suppose the method of cylindrical shells is used to find the volume the volume of the solid obtained when the region bounded by the lines $y = x$, $y = 2x$, $x = 1$, and $x = 2$ is rotated about the y-axis. What is the largest circumference of a cylindrical shell?

    A) 2π B) 6π C) π D) 3π

    E) 5π F) 4π G) 9π H) 8π

    Answer: 4π easy
    _____

8. The region bounded by the curves $y = x - x^2$ and $y = 0$ is rotated about the y-axis. Find the volume of the resulting solid.

    A) $\pi/3$ B) $\pi/4$ C) $\pi/5$ D) $\pi/6$

    E) $\pi/12$ F) $\pi/24$ G) $\pi/30$ H) $\pi/36$

    Answer: $\pi/6$ medium
    _____

9. Find the volume of the solid obtained by rotating about the y-axis the region bounded by

$y = \sqrt{1+x^2}$, $y = 0$, $x = 0$, and $x = 1$.

A) $\frac{\pi}{3}\left(\sqrt{3}-1\right)$

B) $\frac{2\pi}{3}\left(2\sqrt{2}-1\right)$

C) $\frac{4\pi}{3}\left(3\sqrt{3}-1\right)$

D) $\frac{\pi}{4}\left(5\sqrt{5}-1\right)$

E) $\frac{3\pi}{4}\left(5\sqrt{2}-1\right)$

F) $\frac{5\pi}{4}\left(3\sqrt{2}-1\right)$

G) $\frac{\pi}{5}\left(7\sqrt{7}-1\right)$

H) $\frac{4\pi}{5}\left(7\sqrt{5}-1\right)$

Answer: $\frac{2\pi}{3}\left(2\sqrt{2}-1\right)$ medium
_____

10. Write the integral that gives the volume of the solid obtained by rotating the region

$R = \{(x,y) \mid 0 \le y \le f(x), a \le x \le b\}$ about the y-axis.

A) $\displaystyle\int_a^b f(x)\,dx$

B) $\displaystyle\int_a^b \pi\, f(x)\,dx$

C) $\displaystyle\int_a^b 2\pi x\, f(x)\,dx$

D) $\displaystyle\int_a^b 2\pi y\, f(x)\,dx$

E) $\displaystyle\int_a^b \left[f(x)\right]^2 dx$

F) $\displaystyle\int_a^b \pi \left[f(x)\right]^2 dx$

G) $\displaystyle\int_a^b \pi x \left[f(x)\right]^2 dx$

H) $\displaystyle\int_a^b x^2 f(x)\,dx$

Answer: $\displaystyle\int_a^b 2\pi x\, f(x)\,dx$ easy
_____

11. Find the volume of the solid obtained by rotating about the y-axis the region bounded by

$y = \cos x$, $y = 0$, $x = 0$, and $x = \pi/2$.

A) π

B) π^2

C) $\pi^2 - (\pi/2)$

D) $\pi^2 - \pi$

E) $\pi^2 - 2\pi$

F) $2\pi - (\pi^2/4)$

G) $2\pi - (\pi^2/2)$

H) $\pi - (\pi^2/4)$

Answer: $\pi^2 - 2\pi$ hard
_____

12. Find the volume of the region obtained by rotating about the y-axis the region bounded by

$y = \sin x$ and $y = 0$ from $x = 0$ to $x = \pi$.

A) $\pi/4$

B) $\pi/2$

C) π

D) 2π

E) $\pi^2/4$

F) $\pi^2/2$

G) π^2

H) $2\pi^2$

Answer: $2\pi^2$ hard
_____

13. The region in the xy-plane bounded by $y = x^2$, $y = 0$ and $x = 1$ is revolved around the y-axis. Find the volume of the solid generated.

Answer: $\frac{\pi}{2}$ medium

14. Given the curves A: $y^2 = 2(x-1)$ and B: $y^2 = 4x(x-2)$. Sketch the curves labeling all points of intersection. Find the volume of the solid of revolution generated by revolving the bounded region about the x-axis. Find the volume of the solid of revolution generated by revolving the bounded region about the y-axis.

Answer:

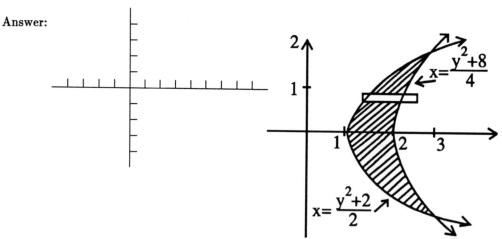

Around x-axis: 2π; around y-axis: $\frac{48\pi}{5}$; medium

15. Use cylindrical shells to find the volume of the solid obtained by revolving around the y-axis the region bounded by the curves $y^2 = 8x$ and $x = 2$.

Answer: $\frac{128\pi}{5}$ medium

16. A cylindrical hole has been drilled straight through the center of a sphere of radius R. Use the method of cylindrical shells to find the volume of the remaining solid if it is 6 cm high.

Answer: 36π medium

17. Let R be the region bounded by the curves: $y = \frac{1}{x}$, $y = x^2$, $x = 0$, $y = 2$. Suppose R is revolved around the x-axis. Set up (but do NOT evaluate) the integral for the volume of rotation using the method of cylindrical shells.

Answer: $2\pi \displaystyle\int_0^1 y^{\frac{3}{2}} \, dy + 2\pi \int_1^2 dy$ medium

18. Find the volume of the solid generated by revolving about the line $x = -1$, the region bounded by the curves $y = -x^2 + 4x - 3$ and $y = 0$.

Answer: 8π medium

19. Find the volume of the solid formed when the region R bounded by $y = x^3 + 1$, $x = 1$ and $y = 1$ is rotated about the line $x = -1$.

Answer: $\frac{9\pi}{10}$ medium

Calculus, 2nd Edition
by James Stewart
Chapter 5, Section 4
Work

1. An anchor weighing 60 pounds is lifted 10 feet. How much work in foot-pounds is done?

    A) 900   B) 600   C) 200   D) 1200

    E) 800   F) 100   G) 60   H) 300

    Answer: 600 easy

2. A spring stretches 1 foot beyond its natural position under a force of 100 pounds. How much work in foot-pounds is done in stretching it 3 feet beyond its natural position?

    A) 600   B) 30   C) 1500   D) 450

    E) 100   F) 900   G) 150   H) 300

    Answer: 450 medium

3. A rope 100 feet long weighing 2 pounds per foot hangs over the edge of a building 100 feet tall. How much work in foot-pounds is done in pulling the rope to the top of the building?

    A) 7500   B) 5000   C) 1000   D) 500

    E) 10000   F) 750   G) 1250   H) 12500

    Answer: 10000 medium

4. A rope 100 feet long weighing 2 pounds per foot hangs over the edge of a building 100 feet tall. How much work in foot-pounds is required to pull 20 feet of the rope to the top of the building?

    A) 4000   B) 3500   C) 3800   D) 4200

    E) 4100   F) 3700   G) 3900   H) 3600

    Answer: 3600 medium

5. A right circular cylinder tank of height 1 foot and radius 1 foot is full of water. Taking the density of water to be a nice round 60 pounds per cubic foot, how much work in foot-pounds is required to pump all of the water up and over the top of the tank?

A) 8π B) 24π C) 16π D) 6π

E) 5π F) 4π G) 30π H) 18π

Answer: 30π medium

---

6. A right circular conical tank of height 1 foot and radius 1 foot at the top is full of water. Taking the density of water to be a nice round 60 pounds per cubic foot, how much work in foot-pounds is required to pump all the water up and over the top of the tank?

A) 24π B) 16π C) 5π D) 8π

E) 6π F) 30π G) 4π H) 18π

Answer: 5π hard

---

7. Assuming adiabatic expansion $(PV^{1.4} = k)$, how much work in foot-pounds is done by a steam engine cylinder starting at a pressure of 200 lb/in^2 and a volume of 25 in^3 and expanding to a volume of 800 in^3 ?

A) 9175   B) 9225   C) 9275   D) 9125

E) 9375   F) 9325   G) 9075   H) 9425

Answer: 9375   hard

---

8. Find the work done in stretching a spring 6 inches beyond its natural length, if the spring constant $k = 20$ lb/ft.

Answer: 2.5 ft-lbs.   easy

---

9. A hemispherical tank with radius 8 feet is filled with water to a depth of 6 feet. Find the work required to empty the tank by pumping the water to the top of the tank.

Answer: 900 πw ft-lbs where w is the number of pounds in the weight of 1 ft^3 of water.

10. Suppose a hemispherical tank of radius 10 feet is filled with a liquid whose density is 62 pounds per cubic foot. Find the work required to pump all of the liquid out through the top of the tank.

Answer: $155,000 \, \pi$ ft-lbs   medium

11. An atom is moving radially outward from the origin opposite the pull of the force

$F = \dfrac{a}{(r-b)^3} - \dfrac{c}{r^6}$ where a, b, c are constants and r is the radial distance from the origin. How

much work must be done to move the atom from R_1 to R_2?

Answer: $W = \text{Work} = \dfrac{a}{2}\left(\dfrac{1}{(R_1-b)^2} - \dfrac{1}{(R_2-b)^2}\right) + \dfrac{c}{5}\left(\dfrac{1}{R_2{}^5} - \dfrac{1}{R_1{}^2}\right)$ medium

Calculus, 2nd Edition
by James Stewart
Chapter 5, Section 5
Average Value of a Function

1. Find the average value of the function $f(x) = 2 + 3x$ on the interval $[0, 4]$.

    A) 2   B) 4   C) 16   D) 10

    E) 14   F) 8   G) 6   H) 12

    Answer: 8 medium

2. Find the average value of the function $f(x) = x^3$ on the interval $[1, 3]$.

    A) 8   B) 2   C) 14   D) 10

    E) 6   F) 4   G) 16   H) 12

    Answer: 10 medium

3. Find the average value of the function $f(x) = \sin x$ on the interval $[0, \pi]$.

    A) 2π B) 3π C) π D) 4π

    E) $3/\pi$ F) $4/\pi$ G) $1/\pi$ H) $2/\pi$

    Answer: $2/\pi$ medium

4. The density of a rod 4 meters long is $2 + x$ kg/m at a distance of x meters from one end of the rod. Find the average density of the rod.

    A) 3   B) 5/2   C) 3/2   D) 17/4

    E) 7/2   F) 15/4   G) 4   H) 1/2

    Answer: 4 easy

5. The density of a rod 4 meters long is $\sqrt{x}$ kg/m at a distance of x meters from one end of the rod.
    Find the average density of the rod.

    A) 6   B) 4/3   C) 3   D) 1

    E) 2   F) 16/3   G) 4   H) 8/3

    Answer: 4/3 medium

6. Find the average value of the function $f(x) = \sin 3x$ on the interval $[0, \pi]$.

A) 1 B) 3/2 C) 2 D) 5/2

E) $1/\pi$ F) $3/2\pi$ G) $2/3\pi$ H) $5/2\pi$

Answer: $2/3\pi$ medium
_____

7. Find the average value of $f(x) = x^2 - 2x$ on the interval $[0, 3]$.

Answer: 0 easy
_____

8. Find the average value of $f(x) = x^3 - x$ on the interval $[1, 3]$.

Answer: 8 easy
_____

9. Find the average value of $f(x) = \sqrt{x}$ on the interval $[4, 9]$.

Answer: $\frac{38}{15}$ easy
_____

10. The temperature (in °F) in a certain city t hours after 9 a.m. was approximated by the function

$$T(t) = 50 + 14 \sin \frac{\pi t}{12}$$

Find the average temperature during the period from 9 a.m. to 9 p.m..

Answer: 59°F medium
_____

11. The temperature of a metal rod, 5 m long, is $4x$ (in °C) at a distance x meters from one end of the rod. What is the average temperature of the rod?

Answer: 10°C medium
_____

Calculus, 2nd Edition
by James Stewart
Chapter 6, Section 1
Exponential Functions

1. For what value of x is $2^x = (1/2)^x$?

    A) $\sqrt{2}$ B) 2 C) $-\sqrt{2}$ D) -2

    E) 1 F) 0 G) $1/2$ H) -1

    Answer: 0 easy
    _____

2. Find the minimum value of $2^{|x|}$.

    A) $-\sqrt{2}$ B) $1/2$ C) 2 D) -2

    E) 0 F) 1 G) -1 H) $\sqrt{2}$

    Answer: 1 easy
    _____

3. How can $7^{\sqrt{2}}$ be calculated?

    A) by taking a derivative   B) by taking a square root once

    C) by taking an integral   D) by solving a cubic equation

    E) by taking a square root twice   F) by trigonometry

    G) by solving a quadratic equation   H) as a limit of other powers of 7

    Answer: as a limit of other powers of 7 medium
    _____

4. Find the value of the limit $\lim\limits_{x \to \infty} (\pi/4)^x$.

    A) π B) ∞ C) 0 D) -1

    E) 1 F) $\pi/4$ G) $-\pi$ H) $-\infty$

    Answer: 0 medium
    _____

5. Find the value of the limit $\lim\limits_{x \to \infty} 4^{(x+1)/x}$.

    A) ∞ B) $-1/4$ C) $1/4$ D) 0

    E) -4 F) 4 G) 2 H) -2

    Answer: 4 medium
    _____

6. Find the value of the limit $\lim\limits_{x \to 0^+} 4^{(x+1)/x}$.

A) 0   B) -4 C) 1/4   D) -2

E) 4   F) 2   G) ∞ H) $-1/4$

Answer: ∞ medium

7. Find the value of the limit $\lim\limits_{x \to 0^-} 4^{(x+1)/x}$.

A) 4   B) -2 C) ∞ D) $-1/4$

E) 1/4   F) -4 G) 2   H) 0

Answer: 0   medium

8. Find the value of the limit $\lim\limits_{x \to 3^-} 2^{1/(x-3)}$.

A) $-\infty$ B) -2 C) -1 D) 0

E) 1/2   F) 1   G) 3/2   H) ∞

Answer: 0   medium

9. If $0 < a < 1$, find the value of the limit $\lim\limits_{x \to \infty} a^x$.

A) ∞ B) 1   C) a D) 1/a

E) $\sqrt{a}$ F) a^2 G) 0   H) $-\infty$

Answer: 0   easy

10. Find the value of the limit $\lim\limits_{x \to 2^+} 3^{1/(2-x)}$.

A) $-\infty$ B) -2 C) -1 D) 0

E) 1/2   F) 1   G) 3/2   H) ∞

Answer: 0   medium

11. Make a rough sketch of the graph of $y = (1.1)^x$. Do not use a calculator. Use the basic graphs from Section 6.1 and any needed transformations.

Answer: easy

12. Make a rough sketch of the graph of $y = 2^x + 1$. Do not use a calculator. Use the basic graphs from Section 6.1 and any needed transformations.

Answer: easy

13. Make a rough sketch of the graph of $y = -3^x$. Do not use a calculator. Use the basic graphs from Section 6.1 and any needed transformations.

Answer: medium

14. **Make a rough sketch of the graph of** $y = 2^{|x|}$. **Do not use a calculator. Use the basic graphs from Section 6.1 and any needed transformations.**

Answer: 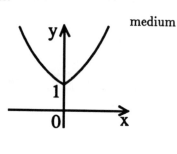   medium

15. **Make a rough sketch of the graph of** $y = 2 + 5\left(1 - 10^{-x}\right)$. **Do not use a calculator. Use the basic graphs from Section 6.1 and any needed transformations.**

Answer: 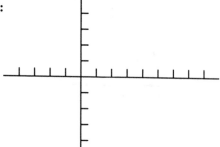   medium

16. **Find the limit:** $\lim\limits_{x \to -\infty} (1.1)^x$.

Answer: 0   easy
_____

17. **Find the limit:** $\lim\limits_{x \to -\infty} \pi^{-x}$.

Answer: ∞ medium
_____

18. **Find the limit:** $\lim\limits_{x \to -(\pi/2)^+} 2^{\tan x}$.

Answer: 0   medium
_____

19. Find the limit: $\lim\limits_{x \to -\infty} 3^{\frac{1}{x}}$.

Answer: 1   medium

20. Find the limit: $\lim\limits_{x \to 0^-} 3^{\frac{1}{x}}$.

Answer: 0   medium

1.   Let $f(x) = e^{-x}$. Find the value of $f'(1)$.

   A) $2e$ B) e C) $2e^{-1}$ D) $-2e$

   E) e^{-1} F) $-e$ G) $-2e^{-1}$ H) $-e^{-1}$

   Answer: $-e^{-1}$ easy

2.   Let $f(x) = e^{-x^2}$. Find the value of $f'(1)$.

   A) $2e^{-1}$ B) $2e$ C) $-2e$ D) $-e$

   E) e F) $-2e^{-1}$ G) e^{-1} H) $-e^{-1}$

   Answer: $-2e^{-1}$ medium

3.   Let $f(x) = e^{x^2}$. Find the value of $f''(0)$.

   A) e B) 1 C) -1 D) e^{-1}

   E) -2 F) $2e^{-1}$ G) $2e$ H) 2

   Answer: 2 medium

4.   Let $f(x) = e^{\ln 2x}$. Find the value of $f'(\pi)$.

   A) 1 B) 2 C) 2π D) -1

   E) -2π F) 0 G) -2 H) π

   Answer: 2 medium

5.   Let $f(x) = e^{e^x}$. Find the value of $f'(1)$.

   A) e^{e+2} B) e^e C) e^{e-1} D) e^{e-2}

   E) e F) e^{e+1} G) e^2 H) $2e^2$

   Answer: e^{e+1} hard

6. Find the value of the integral $\int_0^1 2^x \, dx$.

A) $e - 1$ B) $e^2 - 1$ C) $e^2 - e$ D) 2

E) $1/\ln 2$ F) 4   G) 1   H) $\ln 2$

Answer: $1/\ln 2$ medium
_____

7. Find the value of the integral $\int_0^1 x e^{-x^2} \, dx$.

A) $-e/2$ B) $e/2$ C) $\left(1 - e^{-1}\right)/2$ D) e

E) $-e$ F) $-e^{-1}$ G) $e^{-1}/2$ H) e^{-1}

Answer: $\left(1 - e^{-1}\right)/2$ medium
_____

8. How many points of inflection does $f(x) = x^3 e^{-x}$ have?

A) 0   B) 1   C) 2   D) 3

E) 4   F) 5   G) 6   H) infinitely many

Answer: 3   hard
_____

9. Find the absolute maximum of the function $f(x) = \dfrac{e^{-x}}{1 + x^2}$.

A) $4/3$ B) 1   C) e^{-1} D) $e^{-1}/2$

E) 2   F) $1/2$ G) $3/2$ H) no absolute maximum

Answer: no absolute maximum   (medium)
_____

10. Find the interval on which $f(x) = x e^{-x}$ is increasing.

A) $(-\infty, 1]$ B) $(-\infty, 2]$ C) $(-\infty, 3]$ D) $[1, 2]$

E) $[1, e]$ F) $(-\infty, e]$ G) $(-\infty, \infty)$ H) $(-\infty, 1/e)$

Answer: $(-\infty, 1]$ medium
_____

11. Find the interval on which $f(x) = x e^{-x}$ is concave upward.

A) $(-\infty, 0)$ B) $(-\infty, 1)$ C) $(-\infty, 2)$ D) $(0, 1)$

E) $(1, 2)$ F) $(1, \infty)$ G) $(2, \infty)$ H) $(-\infty, \infty)$

Answer: $(2, \infty)$ medium
_____

12. Find the interval on which the function $f(x) = e^x/x$ is increasing.

A) $(-\infty, -1]$ B) $(-\infty, 0)$ C) $[-1, 1]$ D) $(0, 1/e]$

E) $[1/e, 1]$ F) $(0, e]$ G) $[1/e, \infty)$ H) $[1, \infty)$

Answer: $[1, \infty)$ medium

13. On the domain $(0, \infty)$ find the minimum value of $f(x) = e^x/x$.

A) 0 B) $1/e^2$ C) $1/e$ D) 1

E) $e - 1$ F) e G) $e^2 - 1$ H) e^2

Answer: e medium

14. Let $f(x) = e^x/x$. Find $f''(x)$.

A) $e^x(x+4)/x^4$ B) $e^x(x^2 - 1)/x^4$

C) $e^x(x^2 + x)/x^4$ D) $e^x(x^2 + 3)/x^4$

E) $e^x(x - 2)/x^3$ F) $e^x(x^2 + 5x)/x^3$

G) $e^x(x^2 - 2x + 2)/x^3$ H) $e^x(x^3 - 4x^2 + 3)/x^3$

Answer: $e^x(x^2 - 2x + 2)/x^3$ medium

15. Find y' if $y = e^{\sqrt{x^3 + 1}}$.

Answer: $\dfrac{3x^2 e^{\sqrt{x^3 + 1}}}{2\sqrt{x^3 + 1}}$ medium

16. Find $\dfrac{dy}{dx}$ if $y = e^{xy} + e^{99}$

Answer: $\dfrac{dy}{dx} = \dfrac{ye^{xy}}{1 - xe^{xy}}$ medium

17. Find y' if $y = xe^{\left(x^2 + 7\right)}$.

Answer: $2x^2 e^{x^2 + 7} + e^{x^2 + 7}$ medium

18.   Evaluate the integral: $\int xe^{x^2}\,dx$.

Answer: $\frac{1}{2}e^{x^2}+C$ easy

19.   Evaluate the integral: $\int \frac{e^{\frac{1}{x}}}{x^2}\,dx$

Answer: $-e^{\frac{1}{x}}+C$ medium

20.   Evaluate the integral: $\int e^x \sin(e^x)\,dx$

Answer: $-\cos(e^x)+C$ easy

21.   Find $f(x)$ if $f''(x)=3e^x+5\sin x$, $f(0)=1$, and $f'(0)=2$.

Answer: $f(x)=3e^x-5\sin x+4x-2$ medium

22.   Find the value of the limit $\lim_{x\to 0^-} e^{\cot x}$.

A) $-\infty$ B) $-e$ C) -1 D) 0

E) 1 F) e G) ∞ H) does not exist

Answer: 0 medium

23.   From the choices below, pick the limit whose value is e.

A) $\lim_{x\to 0}(1+x)^x$ B) $\lim_{x\to 0}(1+x)^{1/x}$ C) $\lim_{x\to 0}\left(1+\frac{1}{x}\right)^x$

D) $\lim_{x\to 0}\left(1+\frac{1}{x}\right)^{1/x}$ E) $\lim_{x\to 0} x^{1+(1/x)}$ F) $\lim_{x\to 0} x^{1/x}$

G) $\lim_{x\to 0} x^x$ H) $\lim_{x\to\infty} x^{\ln x}$

Answer: $\lim_{x\to 0}(1+x)^{1/x}$ medium

24. Find the value of the limit $\lim\limits_{x \to -\infty} \frac{e^x}{x}$.

A) $-\infty$ B) $-e$ C) -1 D) 0

E) e^{-1} F) 1 G) e H) ∞

Answer: 0 medium

_____

25. Find the value of the limit $\lim\limits_{x \to 0^+} \frac{e^x}{x}$.

A) $-\infty$ B) $-e$ C) -1 D) 0

E) e^{-1} F) 1 G) e H) ∞

Answer: ∞ medium

_____

26. Find the value of the limit $\lim\limits_{x \to 0^-} \frac{e^x}{x}$.

A) $-\infty$ B) $-e$ C) -1 D) 0

E) e^{-1} F) 1 G) e H) ∞

Answer: $-\infty$ medium

_____

27. Find the value of the limit $\lim\limits_{x \to 0} (1+x)^{1/x}$.

A) 1 B) $1/e$ C) $e-1$ D) e^2-1

E) $e+1$ F) e G) $1/(e-1)$ H) $1/(e+1)$

Answer: e easy

_____

Calculus, 2nd Edition
by James Stewart
Chapter 6, Section 3
Inverse Functions

1   Find the inverse function for $f(x) = 2x + 5$.

A) $x + (5/2)$ B) $x - (5/2)$ C) $(x/2) + 5$ D) $(x+5)/2$

E) $2x - 5$ F) $(x-5)/2$ G) $2x + 5$ H) $(x/2) - 5$

Answer: $(x-5)/2$ easy

2.   Find the inverse function for $f(x) = \frac{x-1}{x+1}$.

A) $(x+1)/(x-1)$ B) $x/(x+1)$ C) $(x+1)/x$

D) $(1+x)/(1-x)$ E) $(x+1)/(x-1)$ F) $x/(x-1)$

G) $(x-1)/(x+1)$ H) $(x-1)/x$

Answer: $(1+x)/(1-x)$ medium

3.   Find the domain of the inverse function for $f(x) = \sqrt{3 + 7x}$.

A) $[0, \infty)$ B) $[7/3, \infty)$ C) $(-\infty, 0]$ D) $[-3/7, \infty)$

E) $[-7/3, \infty)$ F) $(-\infty, 7/3]$ G) $(-\infty, 3/7]$ H) $(-\infty, -7/3]$

Answer: $[0, \infty)$ medium

4.   Find the range of the inverse function for $f(x) = \sqrt{3 + 7x}$.

A) $(-\infty, 0]$ B) $(-\infty, 3/7]$ C) $[0, \infty)$

D) $[7/3, \infty)$ E) $(-\infty, -7/3]$ F) $(-\infty, 7/3]$

G) $[-3/7, \infty)$ H) $[-7/3, \infty)$

Answer: $[-3/7, \infty)$ medium

5.   If the function $f(x) = 3x + 4$ has domain $[2, \infty)$, what is the range of its inverse?

A) $[0, \infty)$ B) $[-2/3, \infty)$ C) $[-3/2, \infty)$

D) $[-3/4, \infty)$ E) $[-4/3, \infty)$ F) $[10, \infty)$

G) $[2, \infty)$ H) $[4, \infty)$

Answer: $[2, \infty)$ easy

6. If the function $f(x) = 3x + 4$ has domain $[2, \infty)$, what is the domain of its inverse?

A) $[-3/2, \infty)$ B) $[2, \infty)$ C) $[-3/4, \infty)$

D) $[10, \infty)$ E) $[-2/3, \infty)$ F) $[-4/3, \infty)$

G) $[4, \infty)$ H) $[0, \infty)$

Answer: $[10, \infty)$ medium

---

7. Given the function $\sin x$ with domain $[-\pi/2, \pi/2]$, find the domain of its inverse.

A) $[-\sqrt{3}/2, \sqrt{3}/2]$ B) $[0, \infty)$ C) $[-\pi, \pi]$

D) $[-1, 1]$ E) $[-\pi/2, \pi/2]$ F) $[-1/2, 1/2]$

G) $(-\infty, \infty)$ H) $[-1/\sqrt{2}, 1/\sqrt{2}]$

Answer: $[-1, 1]$ easy

---

8. Let $f(x) = x + \cos x$ and let g be the inverse function of f. Find the value of $g'(1)$.

A) 0 B) 1/2 C) 1 D) 3/2

E) 2 F) 5/2 G) 3 H) 7/2

Answer: 1 hard

---

9. Suppose g is the inverse function of f and $f(4) = 5$ and $f'(4) = 2/3$. Find the value of $g'(5)$.

A) 1/2 B) 1 C) 3/2 D) 2

E) 5/2 F) 3 G) 7/2 H) 4

Answer: $3/2$ easy

---

10. Suppose g is the inverse function of a $1-1$ differentiable function f and let $G(x) = 1/g(x)$. If $f(3) = 2$ and $f'(3) = 1/9$, find the value of $G'(2)$.

A) 9 B) -1 C) 1/9 D) -9

E) 1 F) 0 G) 6 H) $-1/9$

Answer: -1 hard

---

11. Find the inverse of $f(x) = x^2 - 2x + 3 (x \geq 1)$.

Answer: $f^{-1}(x) = 1 + \sqrt{x - 2}$, $y \geq 1$ medium

---

12.   Find the inverse of the function $f(x) = \sqrt{x-5}$. State the domain of the inverse.

Answer: $f^{-1}(x) = x^2 + 5$, the domain is $[0, \infty)$.   easy

13.   Suppose $G(x) = \displaystyle\int_1^x \sqrt{t^2 + 3}\, dt$. Does the function G have an inverse function? Justify your answer.

Answer: By the Fundamental Theorem, $G'(x) = \sqrt{x^2 + 3}$. So, $G'(x) > 0$ for all x.

Therefore, G is an increasing function and G must have an inverse function.

medium

14.   If $f(x) = x^2 - x - 6$, $x \geq 1$, find $(f^{-1})'(6)$.

Answer: $\frac{1}{7}$ medium

15.   Sketch the graph of f for $f(x) = \sqrt[3]{x}$ and determine if f^{-1} exists. If so, find a formula for $y = f^{-1}(x)$ and also sketch the graph of f^{-1}.

Answer:

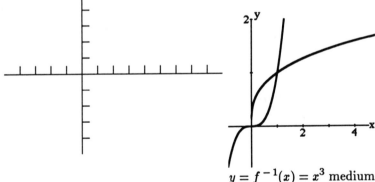

$y = f^{-1}(x) = x^3$ medium

16.   Determine whether or not the given function is one-to-one: $f(x) = x^2 - 2x + 5$.

Answer: not one-to-one   easy

17. Determine whether or not the given function is one-to-one: $g(x) = |x|$.

Answer: not one-to-one   easy

18. Determine whether or not the given function is one-to-one: $h(x) = x^4 + 5$, $0 \leq x \leq 2$.

Answer: one-to-one   easy

19. Find the inverse of $f(x) = \frac{x-2}{x+2}$.

Answer: $f^{-1}(x) = \frac{2(1+x)}{1-x}$ easy

Calculus, 2nd Edition
by James Stewart
Chapter 6, Section 4
Logarithmic Functions

1.   Find the value of $\log_2 1$.

   A) -1 B) $-1/2$ C) 0 D) 10^2

   E) 1 F) $1/2$ G) 2 H) -2

   Answer: 0 easy
   _____

2.   Find the value of $\log_2 16$.

   A) $1/8$ B) 3 C) 2 D) 0

   E) $1/4$ F) 4 G) 1 H) $1/2$

   Answer: 4 easy
   _____

3.   Find the value of $\log_{16} 2$.

   A) $1/2$ B) 2 C) 4 D) 1

   E) 3 F) 0 G) $1/8$ H) $1/4$

   Answer: $1/4$ medium
   _____

4.   Find the value of $\ln e$.

   A) -1 B) $1/\sqrt{e}$ C) e D) 0

   E) $\sqrt{e}$ F) $1/e$ G) 1 H) $-e$

   Answer: 1 easy
   _____

5.   Find the value of $\ln \sqrt{e^3}$.

   A) $2/3$ B) $\sqrt{e}$ C) $e^3/2$ D) $3/2$

   E) e^3 F) $e^3 - 2$ G) $2e/3$ H) $2/e^3$

   Answer: $3/2$ medium
   _____

6.   Find the value of $e^{\ln 8}$.

A) 3   B) 8   C) 1   D) 2

E) 1/8   F) 1/3   G) e^3 H) 4

Answer: 8 easy
_____

7.   Find the value of $\log_2 e - \log_2 (e/16)$.

A) -2 B) e^{-2} C) 4   D) e^{16}

E) -4 F) e^2 G) 2   H) e^{-16}

Answer: 4 medium
_____

8.   Find the value of the limit $\displaystyle\lim_{x \to \infty} \frac{e^x}{e^x + 10}$.

A) 1   B) ln 10   C) 1/ln 10   D) 10

E) 1/10   F) 0   G) ln (1/10)   H) -1

Answer: 1 medium
_____

9.   Find the value of the limit $\displaystyle\lim_{x \to \infty} \frac{\ln x}{\ln \sqrt{x} + 10}$.

A) 1/2   B) ∞ C) 1/11   D) 0

E) 1/10   F) ln 2   G) 2   H) 1

Answer: 2 medium
_____

10. Solve the equation $\log_2(\ln x) = 1$.

A) 2^e B) $2e$ C) $e/2$ D) 1

E) $\sqrt{e}$ F) $1/e$ G) $2/e$ H) e^2

Answer: e^2 medium
_____

11. Solve the equation $e^{2x-2} = 4$.

A) ln 2   B) $1 - \ln 2$ C) $1 + \ln 2$ D) $1 - 2\ln 2$

E) $1 + 2\ln 2$ F) $2 + \ln 2$ G) $2 - \ln 2$ H) $2 - 2\ln 2$

Answer: $1 + \ln 2$ medium
_____

12. Find the value of the limit $\lim\limits_{x \to 0^+} \frac{\ln x}{x}$.

A) $-\infty$ B) -1 C) 0 D) $1/e$

E) 1 F) e G) $e^{1/e}$ H) ∞

Answer: $-\infty$ medium

13. Solve the equation $e^{2x-4} = 16$.

A) $\ln 2$ B) $-\ln 2$ C) $1 + \ln 2$ D) $1 - \ln 2$

E) $1 + 2\ln 2$ F) $1 - 2\ln 2$ G) $2 + 2\ln 2$ H) $2 - 2\ln 2$

Answer: $2 + 2\ln 2$ medium

14. Find the value of the limit $\lim\limits_{x \to 1^+} e^{1/(x-1)}$.

A) $-\infty$ B) -2 C) -1 D) 0

E) $1/2$ F) 1 G) $3/2$ H) ∞

Answer: ∞ medium

15. Find the value of the limit $\lim\limits_{x \to 0^-} \dfrac{1}{1 + e^{1/x}}$.

A) 1 B) $1/2$ C) 0 D) ∞

E) $-\infty$ F) -1 G) $-1/2$ H) 2

Answer: 1 medium

16. Solve the equation $e^{x-2} = 4$.

A) $\ln 2$ B) $-\ln 2$ C) $1 + \ln 2$ D) $1 - \ln 2$

E) $1 + 2\ln 2$ F) $1 - 2\ln 2$ G) $2 + 2\ln 2$ H) $2 - 2\ln 2$

Answer: $2 + 2\ln 2$ medium

17. Find the domain of the function $f(x) = \ln(\ln(\ln x))$.

A) $(0, \infty)$ B) $[1, \infty)$ C) $(1, \infty)$ D) $[e, \infty)$

E) (e, ∞) F) $(-\infty, 0)$ G) $(-\infty, 1)$ H) $(-\infty, e)$

Answer: (e, ∞) medium

18. Solve the equation $e^{x-1} = 4$.

A) $\ln 2$ B) $-\ln 2$ C) $1 + \ln 2$ D) $1 - \ln 2$

E) $1 + 2 \ln 2$ F) $1 - 2 \ln 2$ G) $2 + 2 \ln 2$ H) $2 - 2 \ln 2$

Answer: $1 + 2 \ln 2$ medium

19. Find the value of $\log_{32} 8$.

Answer: $\frac{3}{5}$ easy

20. Find the value of $25^{\log_5 10}$.

Answer: 100 easy

21. Suppose $pH = -\log[H^+]$. Suppose further that for vinegar, the hydrogen ion concentration in moles per liter, is given by $[H^+] = 5.2(10^{-4})$. Find the pH of the vinegar.

Answer: $pH = 3.28$ medium

22. Express $\log_2 x + 5 \log_2(x+1) + \frac{1}{2} \log_2(x-1)$ as a single logarithm.

Answer: $\log_2[x(x+1)^5 \sqrt{x-1}]$ easy

23. Express $\ln x + a \ln y - b \ln z$ as a single logarithm.

Answer: $\ln\left(\frac{xy^a}{z^b}\right)$ easy

24. **Make a rough sketch of the graph of** $y = 1 + \log_5(x-1)$. **Do not use a calculator. Just use the graphs given in your text and any needed transformations.**

Answer:

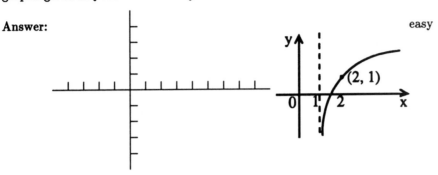

easy

25. **Make a rough sketch of the graph of** $y = \ln\left(\frac{1}{x}\right)$. **Do not use a calculator. Just use the graphs given in your text and any needed transformations.**

Answer:

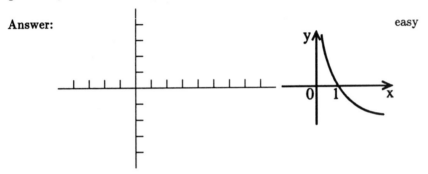

easy

26. Solve for x: $5^{\log_5(2x)} = 6$.

Answer: $x = 3$ easy

27. Solve for x: $\ln x^2 = 2 \ln 4 - 4 \ln 2$.

Answer: $x = \pm 1$ easy

28. A sound so faint that it can just be heard has intensity $I_0 = 10^{-12}$ watt/m^2 at a frequency of 1000 hertz (Hz). The loudness, in decibels (dB), of a sound with intensity I is then defined to be $L = 10 \log_{10}(I/I_0)$. Rock music with amplifiers is measured at 120 dB. The noise from a power mower is measured at 106 dB. Find the ratio of the intensity of the rock music to that of the power mower.

Answer: $10^{1.4} \doteq 25$ medium

29.   Find $f^{-1}(x)$ for $f(x) = \sqrt{e^x + 2}$.

Answer: $f^{-1}(x) = \ln\left(x^2 - 2\right)$ medium

1. Let $f(x) = \log_2 x$. Find the value of $f'(1)$.

   A) 2   B) e^2 C) $\ln(1/2)$ D) $e^{-1/2}$

   E) 2^e F) $1/2$ G) $1/\ln 2$ H) $e^{1/2}$

   Answer: $1/\ln 2$ medium

2. Let $f(x) = \ln(x^2)$. Find the value of $f'(1)$.

   A) 2^e B) e^2 C) $\ln(1/2)$ D) 2

   E) $e^{-1/2}$ F) $1/\ln 2$ G) $1/2$ H) $e^{1/2}$

   Answer: 2   medium

3. Let $f(x) = \ln(\ln(x))$. Find the value of $f'(e)$.

   A) $1/e^2$ B) e^2 C) $1+e$ D) $1/e$

   E) $\ln 2$ F) 1   G) e H) 0

   Answer: $1/e$ medium

4. Let $f(x) = \log_2(3x)$. Find the value of $f'(1)$.

   A) $\ln(1/2)$ B) $e^{-1/3}$ C) e^3 D) 3

   E) $e^{1/3}$ F) 3^e G) $1/3$ H) $1/\ln 2$

   Answer: $1/\ln 2$ hard

5. Let $f(x) = \ln(\ln(x))$. Find the value of $f''(e)$.

   A) $2e^{-1}$ B) e^{-1} C) $-e^{-1}$ D) $-e^{-2}$

   E) $-2e^{-2}$ F) e^{-2} G) $-2e^{-1}$ H) $2e^{-2}$

   Answer: $-2e^{-2}$ hard

6. Find the value of the integral $\int_1^e \frac{1}{x}\, dx$.

   A) 1 B) e C) 1/2 D) -1

   E) -2 F) $-1/2$ G) $\sqrt{e}$ H) 2

   Answer: 1 easy

   _____

7. Find the value of the integral $\int_1^{\sqrt{e}} \frac{1}{x}\, dx$.

   A) 2 B) -2 C) $-1/2$ D) e

   E) $\sqrt{e}$ F) 1 G) 1/2 H) $\sqrt{e}/2$

   Answer: 1/2 medium

   _____

8. Find the value of the integral $\int_0^1 \frac{1}{1+x}\, dx$.

   A) $(1/2)\ln 2$ B) $1/\ln 2$ C) 2 D) $\ln 2$

   E) 1/2 F) 1 G) $2\ln 2$ H) 0

   Answer: $\ln 2$ medium

   _____

9. Find the value of the integral $\int_0^1 \frac{x}{1+x^2}\, dx$.

   A) 1 B) $\frac{1}{2}\ln 2$ C) $2\ln 2$ D) $1/\ln 2$

   E) $\ln 2$ F) 1/2 G) 2 H) 0

   Answer: $\frac{1}{2}\ln 2$ medium

   _____

10. Find the value of $\int_e^{e^2} \frac{\ln x}{x}\, dx$.

    A) $\ln 2$ B) $\frac{1}{2}\ln 2$ C) 1/2 D) 3/2

    E) 1 F) $1/\ln 2$ G) 0 H) $2\ln 2$

    Answer: 3/2 medium

    _____

11. Let $f(x) = \frac{\ln x}{x}$. Find the interval on which f is decreasing.

    A) $(0, 1/e]$ B) $(0, 1]$ C) $(0, 2]$ D) $(1/e, 1]$

    E) $[1/e, 2]$ F) $[1/e, e]$ G) $[1/e, \infty)$ H) $[e, \infty)$

    Answer: $[e, \infty)$ medium

    _____

12. Let $f(x) = \frac{\ln x}{x}$. Find the maximum value of f.

A) 0 B) $1/e$ C) 1 D) e

E) $e^{1/e}$ F) e^e G) e^{-e} H) e^2

Answer: $1/e$ medium

_____

13. Let $f(x) = \frac{\ln x}{x}$. Find the interval on which f is concave upward.

A) $(0, 1/e]$ B) $(0, 1)$ C) $(0, e)$ D) $(1/e, e)$

E) $(1, e)$ F) $(\sqrt{e}, \infty)$ G) (e, ∞) H) $\left(e^{3/2}, \infty\right)$

Answer: $\left(e^{3/2}, \infty\right)$ hard

_____

14. Find the value of the integral $\displaystyle\int_{-e^2}^{-e} \frac{3}{x}\, dx$.

A) -3 B) $-e$ C) e^3 D) $1/3$

E) $\ln 2$ F) $-\ln 3$ G) $e - \ln 3$ H) $e^2 - \ln 3$

Answer: -3 medium

_____

15. Let $f(x) = x\ln(x^2 - 3)$. Find the value of $f'(2)$.

A) 0 B) 2 C) 4 D) 6

E) 8 F) 10 G) 12 H) 14

Answer: 8 medium

_____

16. Find the value of the integral $\displaystyle\int_{e}^{e^4} \frac{dx}{x\sqrt{\ln x}}$.

A) 0 B) 1 C) 2 D) 3

E) 4 F) 5 G) 6 H) 7

Answer: 2 medium

_____

17. Let $f(x) = 2x + \ln x$ and let g be the inverse function of f. Find the value of $g'(2)$.

A) 1 B) $1/2$ C) $1/3$ D) $1/4$

E) $1/\ln 2$ F) $1/(2\ln 2)$ G) $1/e$ H) $1/e^2$

Answer: $1/3$ hard

_____

18. Let $f(x) = \sqrt{x} \ln x$. Find the interval on which f is increasing.

A) $(0, \infty)$ B) $(0, 1]$ C) $[1, \infty)$ D) $(0, e^{-1}]$

E) $[e^{-1}, \infty)$ F) $(0, e^{-2}]$ G) $[e^{-2}, \infty)$ H) $(0, \sqrt{e})$

Answer: $[e^{-2}, \infty)$ medium
_____

19. Let $f(x) = \sqrt{x} \ln x$. Find the interval on which f is concave upward.

A) $(0, \infty)$ B) $(0, 1)$ C) $(1, \infty)$ D) $(0, e^{-1})$

E) (e^{-1}, ∞) F) $(0, e^{-2})$ G) (e^{-2}, ∞) H) (e, ∞)

Answer: $(0, 1)$ hard
_____

20. Find the minimum value of the function $f(x) = x \ln x$.

A) $-e$ B) -1 C) $-1/e$ D) 0

E) e F) $e^{1/e}$ G) e^e H) e^{-e}

Answer: $-1/e$ medium
_____

21. Let $f(x) = \ln(\ln(\ln x))$. Find the value of $f'(e^2)$.

A) $1/\ln 2$ B) $1/\ln 4$ C) $1/\ln 8$ D) $1/e$

E) $1/e^2$ F) $1/(e^2 \ln 2)$ G) $1/(e^2 \ln 4)$ H) $1/(e^2 \ln 8)$

Answer: $1/(e^2 \ln 4)$ medium
_____

22. Find the interval on which the graph of $f(x) = \ln(x^2 + 1)$ is concave upward.

A) $(-1, 1)$ B) $(-1, 2)$ C) $(-2, 1)$ D) $(-2, 2)$

E) $(-1, 3)$ F) $(-3, 2)$ G) $(-3, 3)$ H) $(-\infty, \infty)$

Answer: $(-1, 1)$ medium
_____

23. Find the range of the function $e^{\sin x}$.

A) $[-1, 1]$ B) $\left[e^{-\pi/2}, e^{\pi/2}\right]$ C) $[0, e]$

D) $[0, 1]$ E) $[-e, e]$ F) $\left[e^{-\pi}, e^{\pi}\right]$

G) $\left[e^{-1}, e\right]$ H) $\left[0, e^{\pi/2}\right]$

Answer: $\left[e^{-1}, e\right]$ easy
_____

24. Use logarithmic differentiation to find $\frac{dy}{dx}$ if $y = x^{e^x}$.

Answer: $x^{e^x}(e^x)\left(\frac{1}{x} + \ln x\right)$ medium

25. Find y' if $y = \ln\sqrt{\frac{4x-7}{x^2+2x}}$.

Answer: $\frac{2}{4x-7} - \frac{x+1}{x^2+2x}$ medium

26. If $y = \frac{(x+3)(x^2+1)^3(x+1)^2}{(x^2+10)^{1/2}}$, find y' by logarithmic differentiation.

Answer: $\left(\frac{1}{x+3} + \frac{6x}{x^2+1} + \frac{2}{x+1} - \frac{x}{x^2+10}\right) \cdot \frac{(x+3)(x^2+1)^3(x+1)^2}{(x^2+10)^{1/2}}$ hard

27. Find $\frac{dy}{dx}$ if $y = \ln\left(\frac{\tan x}{x^2+1}\right)$.

Answer: $\frac{\sec^2 x}{\tan x} - \frac{2x}{x^2+1}$ medium

28. Find y' if $y = (\sqrt{x})^x$, $x > 0$.

Answer: $\frac{(\sqrt{x})^x}{2}(1 + \ln x)$ medium

29. Find $\frac{dy}{dx}$ for $y = 3^{x^2} + e^\pi + (1+x^2)^{\sqrt{2}}$.

Answer: $3^{x^2} + \sqrt{2}(1+x^2)^{(\sqrt{2}-1)}(2x)$ medium

30. Find $\frac{dy}{dx}$ for $y = (1+x^2)^{x^3} + \sin x$.

Answer: $(1+x^2)^{x^3}\left[3x^2 \ln(1+x^2) + \frac{2x^4}{1+x^2}\right] + \cos x$ hard

31.   Find the derivative y' for $y = 3^x$.

Answer: $(\ln 3) \cdot 3^x$ easy

32.   Find the derivative y' for $y = x^x$.

Answer: $(1 + \ln x) \cdot x^x$ medium

33.   Find y' if $y = (\ln x)^{\tan x}$.

Answer: $(\ln x)^{\tan x} \left[\dfrac{\tan x}{x \ln x} + \ln(\ln x) \sec^2 x \right]$ hard

34.   Find the derivative of $(\ln \sec x)^4$.

Answer: $4 \tan x (\ln \sec x)^3$ medium

35.   Differentiate and simplify: $y = \ln\!\left(x^3 - 2x\right)$.

Answer: $y' = \dfrac{3x^2 - 2}{x^3 - 2x}$ medium

36.   Differentiate and simplify: $f(x) = x^{\sin x}$.

Answer: $f'(x) = x^{\sin x}\!\left(\dfrac{1}{x} \sin x + \ln x \cos x\right)$ medium

37.   Find the area under the curve $y = \dfrac{1}{x}$ from $x = 1$ to $x = e^2$.

Answer: 2   medium

38. Let $f(x) = x^{2x}$. Find the value of $f'(1)$.

A) 2   B) $e-1$ C) $e^e - 1$ D) $e+1$

E) e F) e^{e+1} G) e^{e-1} H) e^2

Answer: 2 hard

39. Find the value of the integral $\displaystyle\int_0^1 \frac{e^x}{e^x + 1}\,dx$.

A) $e+1$ B) $\ln(e-1)$ C) $(e-1)/2$ D) $\ln(e+1)$

E) $\frac{1}{2}\ln(e-1)$ F) $(e+1)/2$ G) $\frac{1}{2}\ln(e+1)$ H) $e-1$

Answer: $\ln(e+1)$ medium

40. Let $f(x) = x^{\ln x}$. Find $f'(x)$.

A) $x \ln x$ B) $(\ln x)/x$

C) $2x^{\ln x}(\ln x)/x$ D) $x^{\ln x}(1 + x \ln x)$

E) $x^{\ln x}(1 + (\ln x)/x)$ F) $x^{\ln x}(\ln x + \ln(\ln x))$

G) $x^{\ln x}(x + \ln x)$ H) $x^{\ln x}/\ln x$

Answer: $2x^{\ln x}(\ln x)/x$ hard

41. Let $f(x) = (\sin x)^x$. Find the value of $f'(\pi/2)$.

A) 0   B) 1   C) 2   D) $e^{\pi/2}$

E) e^π F) $\pi/2$ G) π H) 2π

Answer: $e^{\pi/2}$ hard

42. Let $f(x) = \displaystyle\int_2^{e^x} \frac{dt}{\sqrt{\ln t}}$. Find the value of $f'(2)$.

A) e^2 B) $e^2/\sqrt{2}$ C) $1/(4e^2\sqrt{2})$ D) $1/(4\sqrt{2})$

E) $1/\ln 2$ F) $e/\ln 2$ G) $\sqrt{2}/e$ H) does not exist

Answer: $e^2/\sqrt{2}$ medium

43. Find the value of $\displaystyle\int_0^{\ln 2} 3e^{4x}\,dx$.

A) 45/4   B) 11   C) 43/4   D) 21/2

E) 41/4   F) 10   G) 39/4   H) 19/2

Answer: 45/4 medium

44. Let $f(x) = \ln(3x^2 + 1 + e^{-x})$. Find the value of $f'(0)$.

A) -1 B) 0 C) $1/2$ D) $-1/2$

E) e^{-1} F) $\sqrt{e}$ G) $1 + \ln 2$ H) $3 \ln 2$

Answer: $-1/2$ medium

_____

45. Let $f(x) = 2^{3^x}$. Find the value of $f'(1)$.

A) $3 \ln 2$ B) $8 \ln 2$ C) $24 \ln 2$ D) $3 \ln 3$

E) $8 \ln 3$ F) $24 \ln 3$ G) $8 \ln 2 \ln 3$ H) $24 \ln 2 \ln 3$

Answer: $24 \ln 2 \ln 3$ hard

_____

46. Let $f(x) = x^{1/x}$. Find the value of $f'(e)$.

A) 0 B) 1 C) 2 D) 3

E) 4 F) 5 G) 6 H) 7

Answer: 0 medium

_____

47. Let $f(x) = x^x$. Find the value of $f'(2)$.

A) 1 B) 2 C) 4 D) $\ln 2$

E) $1 + \ln 2$ F) $2(1 + \ln 2)$ G) $4(1 + \ln 2)$ H) $4(1 + \ln 4)$

Answer: $4(1 + \ln 2)$ medium

_____

48. Find the absolute minimum of $f(x) = x^x$, $x > 0$.

A) 0 B) 1 C) 4 D) e

E) e^e F) $e^{1/e}$ G) $e^{-1/e}$ H) e^{-e}

Answer: $e^{-1/e}$ medium

_____

49. Let $f(x) = 5^{\tan x}$. Find the value of $f'(\pi/4)$.

A) 1 B) 2 C) 5 D) 10

E) $\ln 5$ F) $2 \ln 5$ G) $5 \ln 5$ H) $10 \ln 5$

Answer: $10 \ln 5$ medium

_____

50. Find an equation of the tangent line to the curve $y = e^{-x}$ which is perpendicular to the line $2x - y = 8$.

A) $x + y = 1$ B) $x + y = 2$ C) $x + y = \ln 2$

D) $x + 2y = 1$ E) $x + 2y = 2$ F) $x + 2y = \ln 2$

G) $x + 2y = 1 + \ln 2$ H) $x + 2y = 2 + \ln 2$

Answer: $x + 2y = 1 + \ln 2$ hard

51. Let $f(x) = x^{\tan x}$. Find the value of $f'(x)$.

A) $x^{\tan x} \ln x$

B) $x^{\tan x} \sec^2 x$

C) $x^{\tan x} \sec^2 x \ln x$

D) $x^{\tan x} \tan x$

E) $x^{\tan x}\left(\sec^2 x \ln x + \tan x\right)$

F) $x^{\tan x}\left(\sec^2 x \ln x + (\tan x/x)\right)$

G) $x^{\tan x}\left(\sec^2 x + \ln x\right)$

H) $x^{\tan x}\left(\sec^2 x + \ln x \tan x\right)$

Answer: $x^{\tan x}\left(\sec^2 x \ln x + (\tan x/x)\right)$ hard

52. Let $f(x) = (\ln x)^x$. Find $f'(x)$.

A) x^{-x}

B) $1/\left(x^{x-1}\right)$

C) $(\ln x)^x \ln x$

D) $(\ln x)^x \ln \ln x$

E) $(\ln x)^{x-1}$

F) $(\ln x)(\ln \ln x)$

G) $(\ln x)^x\left(\ln x + 1\right)$

H) $(\ln x)^x\left(\ln \ln x + (1/\ln x)\right)$

Answer: $(\ln x)^x\left(\ln \ln x + (1/\ln x)\right)$ hard

53. Find an equation of the tangent to the curve $y = e^{2x}$ that is perpendicular to the line $x + 4y = 3$.

A) $x - 4y = \ln 3$

B) $4x - y = 3$

C) $x + 2y = \ln 3$

D) $4x + y = \ln 3$

E) $4x - y = 2 \ln 2 - 2$

F) $4x + y = 3 + 3 \ln 3$

G) $4x - y = 2 + 4 \ln 3$

H) $4x - y = 1 + \ln 2$

Answer: $4x - y = 2 \ln 2 - 2$ hard

54. Let $f(x) = x^{\sqrt{x}}$. Find the value of $f'(4)$.

A) 2   B) 4   C) 8   D) 16

E) $2 + \ln 4$ F) $4 + \ln 2$ G) $8 + 4 \ln 4$ H) $16 = 4 \ln 2$

Answer: $8 + 4 \ln 4$ medium

55. Let R be the region bounded by $y = \ln x$, $y = 0$, and $x = e$. Find the area of R integrating with <u>respect</u> <u>to</u> <u>y</u>.

Answer: 1   medium

56. Mr. Spock has accidentally injected himself with the dangerous drug cordrazine. He quickly calculates that the concentration (y), in parts per million, of the drug in his blood t minutes after the injection is given by:

$$y = e^{-t} - e^{-2t}$$

a) At what time will the concentration reach its maximum value?

b) What will the maximum concentration be?

Answer: a) $t = \ln 2$ b) 0.25 ppm   hard

Calculus, 2nd Edition
by James Stewart
Chapter 6, Section 6
The Logarithm Defined as an Integral

1. Using the alternative approach to logarithmic and exponential functions and the partition $P = \{1, 2, 3, 4\}$, what number can we obtain as a lower estimate (or lower bound) for $\ln 4$?

   A) 1   B) 4/3   C) 17/12   D) 3/2

   E) 11/12   F) 7/6   G) 5/4   H) 13/12

   Answer: 13/12 medium

2. Using the alternative approach to logarithmic and exponential functions and the partition $P = \{1, 2, 3, 4\}$, what number can we obtain as a upper estimate (or upper bound) for $\ln 4$?

   A) 11/6   B) 7/4   C) 9/4   D) 25/12

   E) 13/6   F) 23/12   G) 10/6   H) 2

   Answer: 11/6 medium

3. Find the value of $e^{2\ln 3}$.

   A) 6   B) 12   C) 9   D) e^6

   E) e^{12} F) e^8 G) e^9 H) 8

   Answer: 9 easy

4. Find the value of $\exp(2 \exp(\ln(2 \ln 2)))$.

   A) e B) 2   C) e^6 D) 16

   E) 64   F) e^4 G) 4   H) e^2

   Answer: 16 hard

Calculus, 2nd Edition
by James Stewart
Chapter 6, Section 7
Exponential Growth and Decay

1. The radioactive isotope Bismuth-210 has a half-life of 5 days. How many days does it take for 87.5% of a given amount to decay?

   A) 15   B) 8   C) 10   D) 13

   E) 11   F) 9   G) 12   H) 14

   Answer: 15 medium

2. A bacteria culture starts with 200 bacteria and triples in size every half hour. After 2 hours, how many bacteria are there?

   A) 17800   B) 16200   C) 23500   D) 24000

   E) 19300   F) 14800   G) 15700   H) 21000

   Answer: 16200 medium

3. A bacteria culture starts with 200 bacteria and triples in size every half hour. After 45 minutes, how many bacteria are there?

   A) $600\sqrt{3}$ B) $800 \ln 2$ C) $800 \ln 3$ D) 1800

   E) $500\sqrt{2}$ F) 900   G) $45 \ln 3$ H) $1200 \ln(3/2)$

   Answer: $600\sqrt{3}$ medium

4. A bacteria culture starts with 200 bacteria and in 1 hour contains 400 bacteria. How many hours

   does it take to reach 2000 bacteria?

   A) $\ln 400$ B) $\ln 10$ C) 10   D) $\ln 1600$

   E) $\ln 2000$ F) $\ln 200$ G) 5   H) $(\ln 10)/\ln 2$

   Answer: $(\ln 10)/\ln 2$ hard

5. When a child was born, her grandparents placed $1,000 in a savings account at 10% interest compounded continuously, to be withdrawn at age 20 to help pay for college. How much money is in the account at the time of withdrawal?

A) $1000e$ B) $500e$ C) $500e^2$ D) $2000e^2$

E) $4000e$ F) $2000e$ G) $1000e^2$ H) $4000e^2$

Answer: $1000e^2$ medium
_____

6. Radium has a half-life of 1600 years. How many years does it take for 90% of a given amount of radium to decay?

A) $1600/\ln 5$ B) $1600 \ln 2$ C) $1600(\ln 10)/\ln 2$

D) $1600 \ln 5$ E) $1600 \ln 10$ F) $1600/\ln 2$

G) $1600 \ln 10$ H) $1600(\ln 2)/\ln 10$

Answer: $1600 (\ln 10)/\ln 2$ hard
_____

7. Carbon 14, with a half-life of 5700 years, is used to estimate the age of organic materials. What fraction of the original amount of carbon 14 would an object have if it were 2000 years old?

A) $\exp((-57/20) \ln 2)$ B) $(57/20) \ln 2$ C) $\exp((-20/57) \ln 2)$

D) $(20/57) \ln 2$ E) $\exp((57/20) \ln 2)$ F) $(1/57) \ln 20$

G) $\exp((20/57) \ln 2)$ H) $(1/20) \ln 57$

Answer: $\exp((-20/57) \ln 2)$ hard
_____

8. An object cools at a rate (°C/min) equal to k times the difference between its temperature and the surrounding air. Suppose the object takes 10 minutes to cool from 60° C to 40° C in a room kept at 20° C. Find the value of k.

A) e^{-20} B) $\ln 2$ C) $10e^{-20}$

D) $40 \ln 10$ E) $1/2$ F) $e^{-1/20}$

G) $(1/10) \ln (1/2)$ H) $60 \ln (1/2)$

Answer: $(1/10) \ln (1/2)$ medium
_____

9. An object cools at a rate (°C/min) equal to k times the difference between its temperature and the surrounding air. Suppose the object takes 10 minutes to cool from 60° C to 40° C in a room kept at 20° C. How many minutes would it take for the object to cool down from 60° C to 30° C?

A) ln 30   B) 20   C) 15   D) 2 ln 30

E) ln (3/4)   F) 30   G) $e^{30 \ln 2}$ H) 60

Answer: 20   medium
_____

10. A bacteria culture starts with 1000 bacteria and grows at a rate proportional to its size. After 2 hours there are 3000 bacteria. After how many hours will there be 10,000 bacteria?

A) ln 2   B) ln 10   C) ln 100

D) ln 1000   E) (ln 2)/(ln 3)   F) (ln 10)/(ln 3)

G) (ln 100)/(ln 3)   H) (ln 1000)/(ln 3)

Answer: (ln 100)/(ln 3)   medium
_____

11. A thermometer is taken from a room where the temperature is 20° C to the outdoors where the temperature is 5° C. After 1 minute the thermometer reads 12° C. After how many minutes will the thermometer read 6° C?

A) ln 15   B) ln 7   C) (ln 15)/(ln 7)

D) ln 15 − ln 7   E) 1/(ln 15 − ln 7)   F) ln 15/(ln 15 − ln 7)

G) ln 7/(ln 15 − ln 7)   H) (ln 7)/(ln 15)

Answer: ln 15/(ln 15 − ln 7)   hard
_____

12. A bacteria population grows at a rate proportional to its size. The count was 400 after 2 hours and 25,600 after 6 hours. In how many minutes does the population double?

A) 20   B) 25   C) 30   D) 35

E) 40   F) 45   G) 50   H) 55

Answer: 40   medium
_____

13. An object cools at a rate (in °C/min) equal to 1/10 of the difference between its temperature and the surrounding air. If a room is kept at 20° C and the temperature of the object is 28° C, what is the temperature of the object 5 minutes later?

A) 22

B) 24

C) $20 + 5e^{-1/10}$

D) $20 + 8e^{-1/2}$

E) $20 + 5e^{-4/5}$

F) $20 + 8e^{-1/10}$

G) $28 - 8e^{-1/10}$

H) $28 - 10e^{-1/2}$

Answer: $20 + 8e^{-1/2}$ medium

14. In an experiment, a tissue culture has been subjected to ionizing radiation. It was found that the number A of undamaged cells depends on the exposure time, in hours, according to the following formula:

$$A = A_0 e^{kt}, \ t \geq 0$$

If 5000 cells were present initially and 3000 survived a 2 hour exposure, find the elapsed time of exposure after which only half the original cells survive.

Answer: 2.71 hours   medium

15. A lettuce leaf collected from the salad bar at the college cafeteria contains $\frac{99}{100}$ as much carbon C^{14} as a freshly cut lettuce leaf. How old is it? (Use 5700 years for the half-life of C^{14}.)

Answer: about 83 years old   medium

16. Assume that the rate of growth of a population of fruit flies is proportional to the size of the population at each instant of time. If 100 fruit flies are present initially and 200 are present after 5 days, how many will be present after 10 days?

Answer: 400   easy

17. A population of bacteria is known to grow exponentially. If 4 million are observed initially and 9 million after 2 days, how many will be present after 3 days?

Answer: $13\frac{1}{2}$ million   easy

18. It takes money 20 years to triple at a certain rate of interest. How long does it take for money to double at this rate?

Answer: 12.62 years   medium

19. What annual rate of interest will make an investment of P dollars double in five years if the interest is compounded continuously?

Answer: 13.9%   medium

20. Suppose that the number of bacteria in a culture at time t is given by $x = 5^4 e^{3t}$. Use natural logarithms to solve for t in terms of x.

Answer: $t = \dfrac{\ln x - 4 \ln 5}{3}$ medium

21. In 1970, the Brown County groundhog population was 100. By 1980, there were 900 groundhogs in Brown County. If the rate of population growth of these animals is proportional to the population size, how many groundhogs might one expect to see in 1995?

Answer: 24,300   medium

22. Given $f'(t) = k \cdot f(t)$, $f(0) = 35$ and $f(3) = 945$, find $f(4)$.

Answer: 2835   medium

23. In a certain medical treatment a tracer dye is injected into a human organ to measure its function rate and the rate of change of the amount of dye is proportional to the amount present at any time. If a physician injects 0.5g of dye and 30 minutes later 0.1g remains, how much dye will be present in $1\frac{1}{2}$ hours?

Answer: .004g   medium

24. John deposits $100 in a bank and at the same time Mary deposits $200. If John's bank pays 10% interest compounded continuously and Mary's bank pays 8% interest, how long must John wait till his bank account exceeds Mary's? (Use ln 2 = 0.7)

Answer: 35 years   medium

Calculus, 2nd Edition
by James Stewart
Chapter 6, Section 8
Inverse Trigonometric Functions

1. Find the value of $\cos^{-1}(1/2)$.

    A) $\pi/8$ B) $\pi/6$ C) $\pi/2$ D) $\pi/12$

    E) $\pi/3$ F) π G) $3\pi/4$ H) $\pi/4$

    Answer: $\pi/3$ easy

2. Find the value of $\cot^{-1}\sqrt{3}$.

    A) $\pi/3$ B) $\pi/6$ C) π D) $\pi/4$

    E) $\pi/12$ F) $\pi/2$ G) $\pi/4$ H) $\pi/8$

    Answer: $\pi/6$ medium

3. Let $f(x) = \tan^{-1}x$. Find the value of $f'(1)$.

    A) $-1/2$ B) $1/2$ C) $-1/3$ D) 1

    E) $1/3$ F) $-1/4$ G) $1/4$ H) -1

    Answer: $1/2$ easy

4. Let $f(x) = \sin^{-1}(2x)$. Find the value of $f'(0)$.

    A) $-1/2$ B) 2 C) -2 D) 1

    E) $1/2$ F) -1 G) 0 H) $-1/\sqrt{2}$

    Answer: 2 medium

5. Find the value of $\sin(2\tan^{-1}4)$.

    A) $4/17$ B) $2/17$ C) $8/17$ D) $4/15$

    E) $1/15$ F) $8/15$ G) $1/17$ H) $2/15$

    Answer: $8/17$ medium

6. Find the value of the integral $\int_0^1 \frac{1}{1+x^2}\,dx$.

   A) $\pi/6$ 　　　 B) $\frac{1}{2}\ln 2$ 　　　 C) $\pi/4$ 　　　 D) 1

   E) $\pi/2$ 　　　 F) $\pi/8$ 　　　 G) -1 　　　 H) $-\frac{1}{2}\ln 2$

   Answer: $\pi/4$ easy
   _____

7. Find the value of the integral $\int_0^1 \frac{x}{1+x^2}\,dx$.

   A) $\pi/6$ 　　　 B) $\pi/4$ 　　　 C) $\frac{1}{2}\ln 2$ 　　　 D) $-\frac{1}{2}\ln 2$

   E) 1 　　　 F) $\pi/2$ 　　　 G) -1 　　　 H) $\pi/8$

   Answer: $\frac{1}{2}\ln 2$ medium
   _____

8. Find the value of the integral $\int_0^2 \frac{1}{4+x^2}\,dx$.

   A) $\pi/8$ 　　　 B) 1 　　　 C) $\pi/4$ 　　　 D) -1

   E) $\pi/2$ 　　　 F) $\frac{1}{2}\ln 2$ 　　　 G) $\pi/6$ 　　　 H) $-\frac{1}{2}\ln 2$

   Answer: $\pi/8$ medium
   _____

9. Find the value of the integral $\int_0^{1/2} \frac{1}{1+4x^2}\,dx$.

   A) $\pi/2$ 　　　 B) $\pi/6$ 　　　 C) $\pi/8$ 　　　 D) -1

   E) $-\frac{1}{2}\ln 2$ 　　　 F) $\pi/4$ 　　　 G) 1 　　　 H) $\frac{1}{2}\ln 2$

   Answer: $\pi/8$ medium
   _____

10. Find the value of the integral $\int_0^1 \frac{1}{\sqrt{4-x^2}}\,dx$.

   A) $\pi/8$ 　　　 B) $-\frac{1}{2}\ln 2$ 　　　 C) $\frac{1}{2}\ln 2$ 　　　 D) $\pi/6$

   E) $\pi/4$ 　　　 F) -1 　　　 G) 1 　　　 H) $\pi/2$

   Answer: $\pi/6$ hard
   _____

11. Let $f(x) = \tan^{-1}(x^2+1)$. Find the value of $f'(1)$.

   A) 0 　　　 B) 0.1 　　　 C) 0.2 　　　 D) 0.3

   E) 0.4 　　　 F) 0.5 　　　 G) 0.6 　　　 H) 0.8

   Answer: 0.4 medium
   _____

12. Find the value of the limit $\lim\limits_{x \to -2^+} \tan^{-1}\left(\dfrac{x}{x+2}\right)$.

A) $-\infty$ B) $-\pi$ C) $-\pi/2$ D) $-\pi/4$

E) 0 F) $\pi/4$ G) $\pi/2$ H) π

Answer: $-\pi/2$ medium

---

13. Find the domain of the function $f(x) = \sin^{-1}(3 - 2x)$.

A) $[0, 1]$ B) $[1, 2]$ C) $[0, 2]$ D) $[0, 3]$

E) $[1, 3]$ F) $[2, 3]$ G) $[-3, -1]$ H) $[-3, -2]$

Answer: $[1, 2]$ medium

---

14. Let $f(x) = \sin^{-1}(4/x)$. Find the value of $f'(8)$.

A) $-1/(8\sqrt{3})$ B) $-1/(4\sqrt{3})$ C) $-1/(2\sqrt{3})$ D) $-1/\sqrt{3}$

E) $\pi/8$ F) $\pi/4$ G) $\pi/2$ H) π

Answer: $-1/(8\sqrt{3})$ medium

---

15. Find the value of the limit $\lim\limits_{x \to 1} \tan^{-1}\left(\dfrac{1}{(x-1)^2}\right)$.

A) 0 B) 1 C) 2 D) $\pi/4$

E) $\pi/2$ F) π G) ∞ H) does not exist

Answer: $\pi/2$ easy

---

16. Find the value of the integral $\displaystyle\int_0^1 \dfrac{\tan^{-1}x}{1+x^2}\, dx$.

A) $\pi^2/32$ B) $\pi^2/24$ C) $\pi^2/16$ D) $\pi^2/12$

E) $\pi^2/8$ F) $\pi^2/4$ G) $\pi^2/2$ H) π^2

Answer: $\pi^2/32$ medium

---

17. Find the value of the limit $\lim\limits_{x \to \infty} \tan^{-1}(x - x^2)$.

A) $-\infty$ B) $-\pi/2$ C) 0 D) $\pi/2$

E) 1 F) -1 G) ∞ H) does not exist

Answer: $-\pi/2$ medium

18. For $y = \dfrac{\arcsin x}{\sqrt{1 - x^2}}$, $|x| < 1$, find y''. Simplify your answer.

Answer: $y'' = \dfrac{(1 + 2x^2)(\text{Arcsin } x) + 3x(1 - x^2)^{1/2}}{(1 - x^2)^{5/2}}$ hard

19. Find the value of $\sin\left(2 \cos^{-1}\left(-\frac{7}{8}\right)\right)$ correct to three decimal places.

Answer: -0.847 medium

20. Find the exact value of $\arcsin\left(\sin\left(\frac{9\pi}{7}\right)\right)$.

Answer: $\theta = -\dfrac{2\pi}{7}$ easy

21. Find the exact value of $\tan\left(\arccos\left(-\frac{1}{3}\right)\right)$.

Answer: $\tan \theta = -\sqrt{8}$ easy

22. If $x = \frac{1}{3}\cos\theta$, what is $\tan^2\theta$ in terms of x?

Answer: $\dfrac{1 - 9x^2}{9x^2}$ easy

23. Find the value of $\displaystyle\int_0^{1/\sqrt{3}} \dfrac{dx}{1 + 3x^2}$.

Answer: $\dfrac{\pi}{4\sqrt{3}}$ medium

24. Find $\sin\left(\sin^{-1}3x\right)$.

Answer: $3x$ easy
_____

25. Find $\cos\left(\sin^{-1}3x\right)$.

Answer: $\sqrt{1-9x^2}$ medium
_____

26. Find $\tan\left(\sin^{-1}3x\right)$.

Answer: $\dfrac{3x}{\sqrt{1-9x^2}}$ medium
_____

27. Differentiate $f(t) = \tan^{-1}t$.

Answer: $\dfrac{1}{1+t^2}$ easy
_____

28. Differentiate $f(t) = \tan^{-1}\sqrt{1-t}$.

Answer: $\dfrac{-1}{(4-2t)\sqrt{1-t}}$ medium
_____

Calculus, 2nd Edition
by James Stewart
Chapter 6, Section 9
Hyperbolic Functions

1.   If $\sinh x = 2$, what is the value of $\cosh x$?

    A) $\sqrt{3}$ B) $\sqrt{6}$ C) 1   D) 3

    E) $\sqrt{8}$ F) $\sqrt{5}$ G) 2   H) $\sqrt{2}$

    Answer: $\sqrt{5}$ medium

2.   If $\tanh x = 1/2$, what is the value of $\cosh x$?

    A) $\sqrt{5}/2$ B) $2/\sqrt{5}$ C) $\sqrt{5}$ D) $2\sqrt{5}$

    E) $2\sqrt{3}$ F) $\sqrt{3}$ G) $2/\sqrt{3}$ H) $\sqrt{3}/2$

    Answer: $2/\sqrt{3}$ hard

3.   If $\cosh x = 2$, what is the value of $\cosh 2x$?

    A) 3   B) 7   C) 9   D) 4

    E) 8   F) 6   G) 5   H) 10

    Answer: 7 medium

4.   Let $f(x) = \sinh x$. Find the value of $f'(0)$.

    A) $\sqrt{2}$ B) 1/2   C) 2   D) 1/3

    E) 1   F) $1/\sqrt{2}$ G) 3   H) 0

    Answer: 1 easy

5.   Let $f(x) = \tanh(2x)$. Find the value of $f'(0)$.

    A) 0   B) 1/3   C) 1/2   D) $\sqrt{2}$

    E) 1   F) $1/\sqrt{2}$ G) 3   H) 2

    Answer: 2 medium

6. Find the value of $\tanh^{-1} 1/2$.

A) $\frac{1}{2}\ln 3$ B) $\ln 3$ C) $2/\ln 3$ D) $3/\ln 2$

E) $\ln 2$ F) $\frac{1}{3}\ln 2$ G) $2\ln 3$ H) $3\ln 2$

Answer: $\frac{1}{2}\ln 3$ medium

---

7. Let $f(x) = \sinh^{-1} x$. Find the value of $f'(1)$.

A) $1/\sqrt{3}$ B) $\sqrt{2}/3$ C) $1/\sqrt{2}$ D) $1/\sqrt{5}$

E) $2/\sqrt{3}$ F) $2\sqrt{5}$ G) $3/\sqrt{5}$ H) $3/\sqrt{2}$

Answer: $1/\sqrt{2}$ medium

---

8. Find the value of $\int_{-1}^{1} \cosh x \, dx$.

A) e^2 B) $1/e^2$ C) $e - 1/e$ D) $2e^2$

E) $2/e$ F) $1/e$ G) $2e$ H) $2/e^2$

Answer: $e - 1/e$ medium

---

9. Find the value of the integral $\int_{0}^{1} \frac{1}{4 - x^2} \, dx$.

A) $(1/4)\ln 5$ B) $(1/3)\ln 2$ C) $(1/4)\ln 3$ D) $(1/5)\ln 4$

E) $(1/4)\ln 2$ F) $(1/2)\ln 2$ G) $(1/2)\ln 3$ H) $(1/3)\ln 4$

Answer: $(1/4)\ln 3$ hard

---

10. Find the value of the integral $\int_{0}^{3/4} \frac{1}{\sqrt{x^2 + 1}} \, dx$.

A) $\sqrt{3/4}$ B) 2 C) $\ln 4$ D) $\ln (5/4)$

E) $\sqrt{5/4)}$ F) $\ln 2$ G) $\sqrt{2}$ H) $\ln (3/4)$

Answer: $\ln 2$ hard

---

11. Find the value of $\tanh(\ln 3)$.

A) 0.2 B) 0.4 C) 0.6 D) 0.8

E) 1.0 F) 1.2 G) 1.4 H) 1.6

Answer: 0.8 easy

12. Find the value of the limit $\lim\limits_{x\to\infty} \tanh x$.

A) $-\infty$ B) 0   C) 1   D) 2

E) e F) e^2 G) $\ln \sqrt{2}$ H) ∞

Answer: 1   easy

13. Find a simpler form for $\dfrac{1 + \tanh(\ln x)}{1 - \tanh(\ln x)}$.

A) $\sinh x$ B) $\cosh x$ C) $\operatorname{csch} x$ D) $\operatorname{sech} x$

E) $\coth x$ F) x G) x^2 H) x^3

Answer: x^2 medium

14. Find the value of $\sinh(\ln 2)$.

A) 0   B) 1/4   C) 1/2   D) 3/4

E) 2   F) 5/4   G) 3/2   H) 7/4

Answer: 3/4   easy

15. Find $\lim\limits_{x\to\infty} \tanh x$.

Answer: 1

16. Show that $\sinh[\ln x] = \dfrac{x^2 - 1}{2x}$.

Answer: Since $\sinh(u) = \dfrac{e^u - e^{-u}}{2}$,

$$\sinh[\ln x] = \frac{e^{\ln x} - e^{-\ln x}}{2} = \frac{e^x - e^{\ln\left(1/x\right)}}{2}$$

$$= \frac{x - \frac{1}{x}}{2} = \frac{\frac{x^2 - 1}{x}}{2} = \frac{x^2 - 1}{2x} \qquad \text{hard}$$

17. Find the exact coordinates of the two points of inflection of the graph of $y = \operatorname{sech} x$.

Answer: $\left(\pm \ln(1 + \sqrt{2}), \ \dfrac{\sqrt{2}}{2} \right)$ medium

18. Evaluate $\int \cosh(\ln x)\,dx$.

Answer: $\dfrac{x^2}{4} + \dfrac{1}{2}\ln x + C$ medium

_____

19. Find an explicit formula for $\sinh^{-1} x$.

Answer: $\sinh^{-1} = y = \ln\left(x + \sqrt{1 + x^2}\right)$ medium

_____

20. Find $\dfrac{dy}{dx}$ if $x\tanh^{-1}(y) + \dfrac{y}{x} = 0$.

Answer: $\dfrac{dy}{dx} = \dfrac{\left(1 - y^2\right)\left[y - x^2\tanh^{-1}(y)\right]}{x^3 + x - xy^2}$ hard

_____

21. Find the derivative: $g(x) = e^x\sinh x$.

Answer: $e^x\sinh x + e^x\cosh x$ medium

_____

22. Find the derivative: $y = \cos(\sinh x)$.

Answer: $y' = -\sin(\sinh x)\cosh x$ medium

_____

23. Find the derivative: $y = e^{\tanh x}\cosh(\cosh x)$.

Answer: $y' = e^{\tanh x}\mathrm{sech}^2 x\,\cosh(\cosh x) + e^{\tanh x}\sinh(\cosh x)\sinh x$ medium

_____

24. Evaluate the integral: $\int \mathrm{sech}^2 x\,dx$

Answer: $\tanh x + C$ easy

_____

Calculus, 2nd Edition
by James Stewart
Chapter 6, Section 10
Indeterminate Forms and L'Hospital's Rule

1. Find the value of the limit $\lim\limits_{x \to 0^+} \dfrac{\cos x - 1}{x}$.

    A) 1/4   B) −2   C) ∞   D) −1/2

    E) 2   F) 0   G) 1/2   H) 4

    Answer: 0 easy

2. Find the value of the limit $\lim\limits_{x \to 0^+} \dfrac{\cos x - 1}{x^2}$.

    A) −1/2   B) 1/4   C) 0   D) 1/2

    E) 2   F) ∞   G) 4   H) −2

    Answer: −1/2 medium

3. Find the value of the limit $\lim\limits_{x \to 0^+} \dfrac{1 - \sin^2 x}{x^2}$.

    A) 2   B) −2   C) ∞   D) 1/2

    E) 1/4   F) 4   G) −1/2   H) 0

    Answer: ∞ medium

4. Find the value of the limit $\lim\limits_{x \to 0^+} \dfrac{x}{\sin x + \tan x}$.

    A) −2   B) 0   C) −1/2   D) 1/4

    E) 2   F) ∞   G) 1/2   H) 4

    Answer: 1/2 medium

5. Find the value of the limit $\lim\limits_{x \to 1^+} \dfrac{\ln(2x)}{\ln x}$.

    A) 1/4   B) 2   C) 4   D) 1/2

    E) 0   F) −2   G) ∞   H) −1/2

    Answer: ∞ medium

6. Find the value of the limit $\lim\limits_{x \to 0^+} \frac{e^x - 1}{x}$.

A) 0   B) 4   C) $-1/2$ D) 1/4

E) 1   F) 1/2   G) 2   H) -2

Answer: 1 medium

7. Find the value of the limit $\lim\limits_{x \to 0^+} \frac{e^{x^2} - 1}{x}$.

A) 4   B) 1/4   C) -2 D) ∞

E) 2   F) 0   G) 1/2   H) $-1/2$

Answer: 0 medium

8. Find the value of the limit $\lim\limits_{x \to 0^+} x^{2/x}$.

A) 1/2   B) ∞ C) 1   D) e

E) 0   F) 2   G) $\ln 2$ H) $\sqrt{2}$

Answer: 0 hard

9. Find the value of the limit $\lim\limits_{x \to \infty} x^{2/x}$.

A) 2   B) e C) $\ln 2$ D) 0

E) 1   F) ∞ G) 1/2   H) $\sqrt{2}$

Answer: 1 hard

10. Find the value of the limit $\lim\limits_{x \to \infty} x^{1/\ln x}$.

A) 0   B) 2   C) e D) ∞

E) $\ln 2$ F) 1/2   G) 1   H) $\sqrt{2}$

Answer: e hard

11. Find the value of the limit $\lim\limits_{x \to \infty} \left(1 + \frac{3}{x} + \frac{5}{x^2}\right)^x$.

A) 0   B) 1   C) 3   D) 5

E) e^3 F) e^5 G) $\ln 3$ H) $\ln 5$

Answer: e^3 hard

12. Find the value of the limit $\lim_{x \to \infty} \frac{\ln x}{x}$.

A) $-\infty$ B) -1 C) 0 D) $1/e$

E) 1 F) e G) $e^{1/e}$ H) ∞

Answer: 0 easy
_____

13. Sketch the curve $y = \frac{\ln x}{x}$.

A)

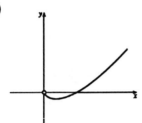

B)

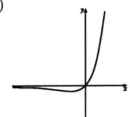

C)

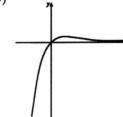

D)

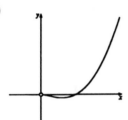

E)

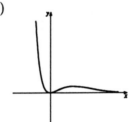

F)

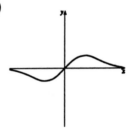

G)

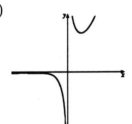

H)

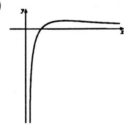

Answer: medium

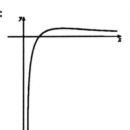

14.   Sketch the curve $y = x \ln x$.

A)

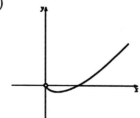

B)

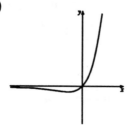

C)

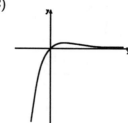

D)

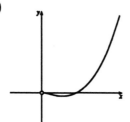

E)

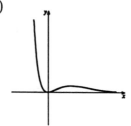

F)

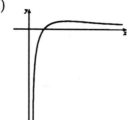

G)

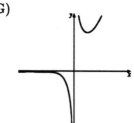

H)

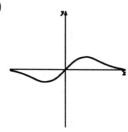

Answer: medium

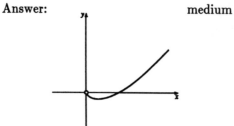

15.   Let $a > 0$ and $b > 0$. Find the value of the limit $\lim\limits_{x \to 0} \dfrac{a^x - b^x}{x}$.

A) 0   B) 1   C) $a - b$ D) $b - a$

E) $\ln (a - b)$ F) $\ln (b - a)$ G) $\ln (a/b)$ H) $\ln (b/a)$

Answer: $\ln (a/b)$ medium

16. Find the value of the limit $\lim_{x \to 1} \left(\frac{1}{\ln x} - \frac{1}{x-1} \right)$.

A) -1 B) $-1/2$ C) 0 D) $1/2$

E) 1 F) ∞ G) $-\infty$ H) does not exist

Answer: $1/2$ medium

17. Let a, b, c, d all be positive constants. Find the value of the limit $\lim_{x \to \infty} \frac{\ln(a + be^{cx})}{dx}$.

A) 0 B) a/d C) b/d D) c/d

E) b F) c G) d H) ∞

Answer: c/d medium

18. Find the value of the limit $\lim_{x \to 0} \frac{1 - \cos \lambda x}{x^2}$.

A) λ B) λ^2 C) $\lambda/2$ D) $\lambda^2/2$

E) $1/2$ F) $1/\lambda$ G) $3/\lambda^2$ H) ∞

Answer: $\lambda^2/2$ medium

19. Find the value of the limit $\lim_{x \to 0} \frac{6^x - 2^x}{x}$.

A) 0 B) 1 C) 2 D) 3

E) 6 F) $\ln 2$ G) $\ln 3$ H) $\ln 4$

Answer: $\ln 3$ medium

20. Find the value of the limit $\lim_{x \to \infty} \left(1 + \frac{2}{x} \right)^{3x}$.

A) e^2 B) e^3 C) e^5 D) e^6

E) $\ln 2$ F) $\ln 3$ G) $\ln 5$ H) $\ln 6$

Answer: e^6 medium

21. Find the value of the limit $\lim_{x \to 0^+} \sqrt{x} \ln x$.

A) $-\infty$ B) -1 C) 0 D) 1

E) e F) ∞ G) $\sqrt{e}$ H) $-\sqrt{e}$

Answer: 0 easy

22. Find the value of the limit $\lim\limits_{x \to 0} \dfrac{e^x - 1 - x - (x^2/2)}{x^3}$.

A) 0 B) 1 C) 1/2 D) 1/3

E) 1/6 F) ∞ G) $-1/2$ H) $-1/3$

Answer: 1/6 medium

23. Sketch the curve $y = \dfrac{e^x}{x}$.

A)
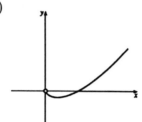

B)
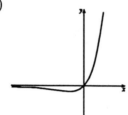

C)
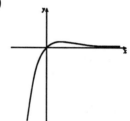

D)
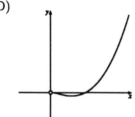

E)
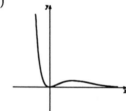

F)
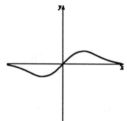

G)
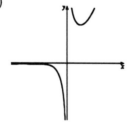

H)
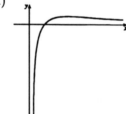

Answer:
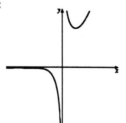

24. Find the value of the limit $\lim_{x \to \infty} \frac{e^x}{x}$.

A) $-\infty$ B) $-e$ C) -1 D) 0

E) e^{-1} F) 1 G) e H) ∞

Answer: ∞ easy

_____

25. Sketch the curve $y = xe^{-x}$.

A)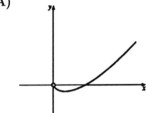

B)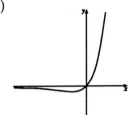

C)

D)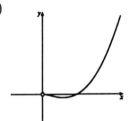

E)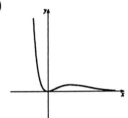

F)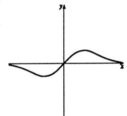

G)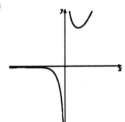

H)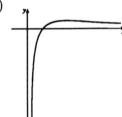

Answer: 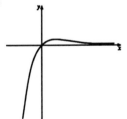   medium

26. Find the value of the limit $\lim\limits_{x \to 0+} \left(\frac{x}{1+x}\right)^x$.

A) 0   B) 1   C) 2   D) 3

E) e F) $1/e$ G) e^2 H) $1/e^2$

Answer: 1   medium

27. Find the value of the limit $\lim\limits_{x \to \infty} xe^{-x}$.

A) -2 B) -1 C) 0   D) 1

E) 2   F) e G) $1/e$ H) ∞

Answer: 0   easy

28. Evaluate: $\lim\limits_{x \to \infty} \dfrac{e^x - e^{2x}}{e^x + e^{-x}}$.

Answer: $-\infty$ medium

29. Evaluate the following limit, if it exists: $\lim\limits_{x \to 0} \dfrac{e^{-2x} - 1}{x^2 - x}$.

Answer: 2   medium

30. Evaluate the limit: $\lim\limits_{x \to 0} \left(\dfrac{\sin 6x}{\sin x}\right)^2$.

Answer: 36   medium

31. Evaluate the limit: $\lim\limits_{x \to 0} \dfrac{e^{5x} - 5x - 1}{x^2}$.

Answer: $\dfrac{25}{2}$ medium

32. Evaluate the limit: $\lim\limits_{x \to 0} \dfrac{2e^x - x^2 - 2x - 2}{x^4 + x^3}$.

Answer: $\frac{1}{3}$ medium

33. Find $\lim\limits_{x \to 0} \dfrac{\ln(1+x)}{x^3}$, if it exists.

Answer: ∞ medium

34. Evaluate the limit: $\lim\limits_{x \to 0} (1 + \sin \pi x)^{1/x}$.

Answer: e^{π} medium

35. Evaluate the limit: $\lim\limits_{x \to 0^+} x^2(\ln x)$.

Answer: 0 medium

36. Sketch the curve $y = x^2 \ln x$.

A)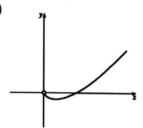

B)

C)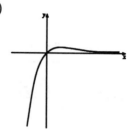

D)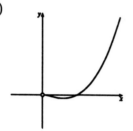

E)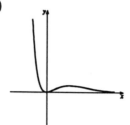

F)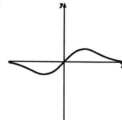

G)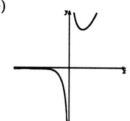

H)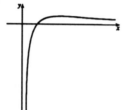

Answer: medium

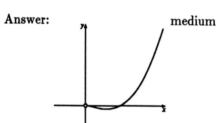

37. Evaluate the limit: $\lim\limits_{x \to \infty} \left(1 - \frac{2}{x}\right)^{3x}$.

Answer: e^{-6} hard

38. Evaluate the limit: $\lim\limits_{x \to 0} \dfrac{xe^{2x} - x}{1 - \cos x}$.

Answer: 4 medium

39. Evaluate the limit: $\lim\limits_{x \to 3} \dfrac{x^n - 3^n}{x^m - 3^m}$, $(m \neq 0)$.

Answer: $\dfrac{n}{m}(3)^{n-m}$ medium

40. Evaluate the limit: $\lim\limits_{x \to 0} \dfrac{e^x - e^{-x}}{\sin x}$.

Answer: 2 medium

41. Evaluate the limit: $\lim\limits_{x \to 0} \dfrac{\arcsin x - x}{\sin x - x}$.

Answer: -1 hard

42. Evaluate the limit if it exists of $f(t) = \dfrac{1 - \cos t}{\sin^2 t}$ as t approaches zero.

Answer: $\dfrac{1}{2}$ easy

43. Evaluate the limit if it exists: $\lim\limits_{x \to 0} \left(\dfrac{1}{x} + \ln x \right)$.

Answer: does not exist medium

44. Evaluate the limit if it exists: $\lim\limits_{x \to 0^+} x^{2x}$.

Answer: 1 medium
_____

45. Evaluate the limit if it exists: $\lim\limits_{x \to \infty} \left(1 + x^2\right)^{1/x}$.

Answer: 1 hard
_____

46. Evaluate the limit if it exists: $\lim\limits_{x \to 1} \dfrac{1 - x + \ln x}{x^3 - 3x + 2}$.

Answer: $-\frac{1}{6}$ medium
_____

47. Evaluate the limit if it exists: $\lim\limits_{x \to \infty} \dfrac{x^3}{e^x}$.

Answer: 0 medium
_____

48. Evaluate the limit if it exists: $\lim\limits_{x \to 0} \left(1 + \sinh x\right)^{3/x}$.

Answer: e^3 hard
_____

49. Evaluate the limit if it exists: $\lim\limits_{x \to 0} \dfrac{x - \sin x}{x^3}$.

Answer: $\frac{1}{6}$ medium
_____

50. Evaluate the limit if it exists: $\lim\limits_{x \to 0} \dfrac{\ln \sec^4 x}{x^2}$.

Answer: 2 medium

Calculus, 2nd Edition
by James Stewart
Chapter 7, Section 1
Integration by Parts

1.   Find the value of the integral $\int_1^e \ln x \, dx$.

   A) 1   B) e^2 C) $e-1$ D) 2

   E) e^2-e F) $e-2$ G) $e-1/2$ H) e

   Answer: 1 medium
   _____

2.   Find the value of the integral $\int_0^1 xe^x \, dx$.

   A) 2   B) e^2-e C) 1   D) e^2

   E) e F) $e-1$ G) $e-2$ H) $(e-1)/2$

   Answer: 1 medium
   _____

3.   Find the value of the integral $\int_1^e \ln x^2 \, dx$.

   A) e^2 B) e C) 2   D) e^2-e

   E) $e-2$ F) 1   G) $e-1$ H) $(e-1)/2$

   Answer: 2 medium
   _____

4.   Find the value of the integral $\int_0^1 xe^{x^2} \, dx$.

   A) 2   B) e C) 1   D) $e-1$

   E) e^2-e F) e^2 G) $(e-1)/2$ H) $e-2$

   Answer: 2 medium
   _____

5.   Find the value of the integral $\int_0^\pi x \sin x \, dx$.

   A) $4\pi-2$ B) $2\pi-2$ C) $\pi/2$ D) $4\pi-4$

   E) $\pi-2$ F) π G) 2π H) 4π

   Answer: π medium
   _____

6. Find the value of the integral $\int_0^{\pi/2} e^x \cos x \, dx$.

A) $(e^{\pi/4}+1)/2$ B) $(e^{\pi/2}+1)/2$ C) $(e^{\pi/4}+1)/4$

D) $(e^{\pi/2}-1)/4$ E) $(e^{\pi/4}-1)/4$ F) $(e^{\pi/4}-1)/2$

G) $(e^{\pi/2}+1)/4$ H) $(e^{\pi/2}-1)/2$

Answer: $(e^{\pi/2}-1)/2$ hard
_____

7. Find the value of the integral $\int_0^1 e^{\sqrt{x}} \, dx$.

A) e B) $2e-1$ C) $2e-2$ D) $2e$

E) $e-2$ F) $e-1$ G) 1 H) 2

Answer: 2 hard
_____

8. Find the value of the integral $\int_0^1 x \tan^{-1} x \, dx$.

A) $\pi/4$ B) $\pi-2$ C) $\pi/2$ D) $(\pi-2)/2$

E) $(\pi-2)/4$ F) $\pi-1$ G) $(\pi-1)/2$ H) $(\pi-1)/4$

Answer: $(\pi-2)/4$ hard
_____

9. Find the value of the integral $\int_0^\pi x \cos x \, dx$.

A) π B) 2π C) 2 D) 0

E) -2 F) 1 G) $1/2$ H) $\pi/2$

Answer: -2 medium
_____

10. Find the value of the integral $\int_0^1 2xe^x \, dx$.

A) 0 B) 1 C) 2 D) 3

E) e F) $e-1$ G) $2e-1$ H) $2e$

Answer: 2 medium
_____

11. Find the value of the integral $\displaystyle\int_1^4 \sqrt{t}\ln t\, dt$.

A) $4\ln 4$

B) $\frac{4}{3}\ln 4$

C) $\frac{8}{3}\ln 4 - \frac{22}{9}$

D) $\frac{16}{3}\ln 4 - \frac{28}{9}$

E) $\frac{22}{3}\ln 4 - 3$

F) $\frac{25}{3}\ln 4 - \frac{35}{9}$

G) $\frac{32}{3}\ln 4 - \frac{16}{9}$

H) $12\ln 4 - \frac{25}{9}$

Answer: $\frac{16}{3}\ln 4 - \frac{28}{9}$ hard

12. Find the value of the integral $\displaystyle\int_0^1 xe^{2x}\, dx$.

A) $1 - e^2$

B) $e + e^2$

C) $(1-e)/2$

D) $(e^2 + 3)/2$

E) $(2 - e^2)/3$

F) $(e^2 + 1)/4$

G) $(e^2 - 3)/4$

H) $(1 - e^2)/8$

Answer: $(e^2 + 1)/4$ medium

13. Find the value of the integral $\displaystyle\int_1^2 x^3 \ln x\, dx$.

A) $\pi/16$

B) $3\pi - \frac{1}{4}$

C) $8\ln 2 - \frac{3}{4}$

D) $2\ln 2 - 1$

E) $4\ln 2 - \frac{15}{16}$

F) $\ln 4 - \frac{7}{8}$

G) $3\ln 4 - \frac{\pi}{16}$

H) $5\ln 4 - \frac{3\pi}{4}$

Answer: $4\ln 2 - \frac{15}{16}$ medium

14. Evaluate the following integral: $\displaystyle\int x\cot^{-1}(x)\, dx$.

Answer: $\frac{1}{2}x^2\cot^{-1}(x) + \frac{1}{2}x - \frac{1}{2}\tan^{-1}(x) + C$ hard

15. Evaluate the following integral: $\displaystyle\int \ln\left[(e^x \ln x)^{1/x}\right] dx$.

Answer: $\ln x\,[\ln(\ln x) - 1]$ hard

16. Determine a reduction formula for $\displaystyle\int_1^e x(\ln x)^n\, dx$.

Answer: $I_n = \dfrac{e^2}{2} - \dfrac{n}{2} I_{n-1}$ where $I_{n-1} = \displaystyle\int_1^e x(\ln x)^{n-1}\, dx$ hard

17. Find $\displaystyle\int x^2 e^{2x}\, dx$.

Answer: $\frac{1}{2} x^2 e^{2x} - \frac{1}{2} x e^{2x} + \frac{1}{4} e^{2x} + C$ hard

18. Evaluate the following integral: $\displaystyle\int (\ln x)^3\, dx$.

Answer: $x(\ln x)^3 - 3x(\ln x)^2 + 6x \ln x - 6x + C$ hard

19. Evaluate the following integral: $\displaystyle\int_1^e (\ln x)^2\, dx$.

Answer: $e - 2$ medium

20. Find the area of the region bounded by the curve $y = \tan^{-1} x$, the x-axis and the line $x = 1$.

Answer: $\dfrac{\pi - \ln 4}{4}$ medium

21. Evaluate the following integral: $\displaystyle\int \dfrac{\ln 2x\, dx}{x}$.

Answer: $\ln 2(\ln x) + \frac{1}{2} (\ln x)^2 + C$ medium

22. Evaluate the following integral: $\displaystyle\int x^2 \cos x\, dx$.

Answer: $x^2 \sin x + 2x \cos x - 2 \sin x + C$ hard

23. Evaluate the following integral: $\int \sec^3 x \, dx$.

Answer: $\dfrac{\sec x \tan x}{2} + \dfrac{\ln |\sec x + \tan x|}{2} + \dfrac{C}{2}$ hard

24. Evaluate the following integral: $\int x^3 e^{3x} \, dx$.

Answer: $\dfrac{1}{3} x^3 e^{3x} - \dfrac{1}{3} x^2 e^{3x} + \dfrac{2}{9} x e^{3x} - \dfrac{2}{27} e^{3x} + C$ hard

25. Evaluate the following integral: $\int x e^{3x} \, dx$.

Answer: $\dfrac{1}{3} x e^{3x} - \dfrac{1}{9} e^{3x} + C$ medium

26. Evaluate the following integral: $\int x^3 e^{x^2} \, dx$.

Answer: $\dfrac{1}{2} x^2 e^{x^2} - \dfrac{1}{2} e^{x^2} + C$ medium

27. Evaluate the following integral: $\int x^2 \cos 2x \, dx$.

Answer: $\dfrac{x^2}{2} \sin 2x + \dfrac{x}{2} \cos 2x - \dfrac{1}{4} \sin 2x + C$ medium

Calculus, 2nd Edition
by James Stewart
Chapter 7, Section 2
Trigonometric Integrals

1. Find the value of the integral $\int_0^\pi \sin^2 x \, dx$.

    A) $2\pi - 4$ B) $\pi - 2$ C) $\pi - 1$ D) π

    E) $\pi/2$ F) 2π G) $2\pi - 2$ H) $(\pi/2) - 1$

    Answer: $\pi/2$ easy

2. Find the value of the integral $\int_0^\pi \sin^3 x \, dx$.

    A) 1 B) $4/3$ C) 0 D) 8π

    E) $2\pi/3$ F) $4\pi/3$ G) $2/3$ H) $8/3$

    Answer: $4/3$ medium

3. Find the value of the integral $\int_0^\pi \sin^2 x \cos x \, dx$.

    A) $4/3$ B) 1 C) 0 D) $2/3$

    E) $4\pi/3$ F) $2\pi/3$ G) $8\pi/3$ H) $8/3$

    Answer: 0 easy

4. Find the value of the integral $\int_0^\pi \cos^2 x \sin x \, dx$.

    A) 0 B) $8\pi/3$ C) $8/3$ D) $4/3$

    E) $2/3$ F) 1 G) $2\pi/3$ H) $4\pi/3$

    Answer: $2/3$ medium

5. Find the value of the integral $\int_0^\pi \sin 2x \cos x \, dx$.

    A) $1/2$ B) $\pi/4$ C) 1 D) $3\pi/4$

    E) $4/3$ F) $2/3$ G) $\pi/2$ H) 0

    Answer: $4/3$ medium

6. Find the value of the integral $\displaystyle\int_0^{\pi/4} \sin 4x\, dx$.

A) 1   B) 0   C) 1/4   D) $\pi/2$

E) $\pi/4$ F) $3\pi/4$ G) 3/4   H) 1/2

Answer: 1/2 easy

7. Find the value of the integral $\displaystyle\int_0^{\pi/4} \sec^2 x\, dx$.

A) 1/2   B) $\pi/4$ C) 3/4   D) $3\pi/4$

E) $\pi/2$ F) 1/4   G) 1   H) 0

Answer: 1 easy

8. Find the value of the integral $\displaystyle\int_0^{\pi/4} \tan x \sec^2 x\, dx$.

A) 1/2   B) $\pi/4$ C) $3\pi/4$ D) 1

E) $\pi/2$ F) 1/4   G) 0   H) 3/4

Answer: 1/2 medium

9. Find the value of the integral $\displaystyle\int_0^{\pi/3} \sec^3 x \tan x\, dx$.

A) 5/3   B) 7/3   C) $\pi^2/27$ D) $\sqrt{3/2}$

E) $3\sqrt{3}/8$ F) 3/7   G) 3/5   H) $\pi^3/27$

Answer: 7/3 medium

10. Find the value of the integral $\displaystyle\int_0^{\pi/4} \tan^2 x \sec^4 x\, dx$.

A) 2/15   B) 4/15   C) 2/5   D) 8/15

E) 2/3   F) 4/5   G) 14/15   H) 16/15

Answer: 8/15 medium

11. Find the value of the integral $\displaystyle\int_0^{\pi/2} \sin^2 x \cos^3 x\, dx$.

A) 2/15   B) 4/15   C) 2/5   D) 8/15

E) 2/3   F) 4/5   G) 14/15   H) divergent

Answer: 2/15 medium

12. Find the integral $\int \sin^3 x \, dx$.

    A) $\frac{\cos^3 x}{3} + C$ B) $\frac{\sin^3 x}{3} + C$ C) $-\cos x + \frac{\cos^3 x}{3} + C$

    D) $\sin x - \frac{\sin^3 x}{3} + C$ E) $\frac{\cos^4 x}{4} + C$ F) $\frac{\sin^4 x}{4} + C$

    G) $\cos x - \frac{\sin^3 x}{3} + C$ H) $\sin x - \frac{\cos^3 x}{3} + C$

Answer: $-\cos x + \frac{\cos^3 x}{3} + C$ easy

13. Find the value of the integral $\int_0^\pi \cos^2 \theta \, d\theta$.

    A) $1/3$ B) $2/3$ C) $1/2$ D) 1

    E) $\pi/3$ F) $2\pi/3$ G) $\pi/2$ H) π

Answer: $\pi/2$ easy

14. Find the integral $\int \frac{\tan^2 x}{\cos^2 x} \, dx$.

Answer: $\frac{1}{3} \tan^3 x + C$ easy

15. Evaluate the following integral: $\int_{\pi/6}^{\pi/2} \cos^3 x \sqrt{\sin x} \, dx$.

Answer: $\dfrac{32\sqrt{2} - 25}{84\sqrt{2}}$ hard

16. Find the following integral: $\int \sin^4 x \cos^3 x \, dx$.

Answer: $\frac{1}{5} \sin^5 x - \frac{1}{7} \sin^7 x + C$ medium

17. Find the following integral: $\int \frac{\sin^5 x + \sin^3 x}{\cos^4 x} \, dx$.

Answer: $\dfrac{-2}{3 \cos^3 x} + \dfrac{3}{\cos x} + \cos x + C$ hard

18. Find $\int \sec^4(3x)\, dx$.

Answer: $\dfrac{\tan(3x)}{3} + \dfrac{\tan^3(3x)}{9} + C$ medium

19. Evaluate $\int \cos(2x)\tan x\, dx$.

Answer: $-\frac{1}{2}\cos(2x) + \ln|\cos x| + C$ medium

20. Find the area under the curve $y = 5\sec^2(2x)$ and above the interval $[\frac{\pi}{8}, \frac{\pi}{6}]$.

Answer: $\frac{5}{2}(\sqrt{3} - 1)$ medium

21. Integrate: $\displaystyle\int \dfrac{dx}{\sin \frac{1}{2} x}$.

Answer: $2\ln|\csc \frac{1}{2} x - \cot \frac{1}{2} x| + C$ medium

22. Integrate: $\displaystyle\int \cos^3 t\, dt$.

Answer: $\sin t - \dfrac{\sin^3 t}{3} + C$ medium

23. Evaluate the integral: $\displaystyle\int \sqrt{\cos x}\, \sin^3 x\, dx$.

Answer: $-\frac{2}{3}(\cos x)^{3/2} + \frac{2}{7}(\cos x)^{7/2} + C$ medium

Calculus, 2nd Edition
by James Stewart
Chapter 7, Section 3
Trigonometric Substitution

1. Find the value of the integral $\int_0^1 \frac{1}{1+x^2}\, dx$.

    A) 1   B) 3/4   C) $\pi/2$ D) $3\pi/4$

    E) 1/4   F) 0   G) $\pi/4$ H) 1/2

    Answer: $\pi/4$ easy

2. Find the value of the integral $\int_0^1 \frac{dx}{\sqrt{x^2+1}}$.

    A) 1   B) $\ln(\sqrt{2}-1)$ C) $\ln(\sqrt{3}+1)$ D) $\ln 2$

    E) 0   F) $\ln(\sqrt{2}+1)$ G) $\ln(\sqrt{3}-1)$ H) $\ln 3$

    Answer: $\ln(\sqrt{2}+1)$ medium

3. Find the value of the integral $\int_0^1 \sqrt{x^2+1}\, dx$.

    A) $\left(\sqrt{2}+\ln(\sqrt{2}-1)\right)/2$ B) $\left(\sqrt{2}-\ln(\sqrt{2}-1)\right)/2$

    C) $\left(\sqrt{2}-\ln(\sqrt{2}+1)\right)/2$ D) $\left(\sqrt{3}+\ln(\sqrt{3}-1)\right)/2$

    E) $\left(\sqrt{3}-\ln(\sqrt{3}+1)\right)/2$ F) $\left(\sqrt{3}+\ln(\sqrt{3}+1)\right)/2$

    G) $\left(\sqrt{3}-\ln(\sqrt{3}-1)\right)/2$ H) $\left(\sqrt{2}+\ln(\sqrt{2}+1)\right)/2$

    Answer: $\left(\sqrt{2}+\ln(\sqrt{2}+1)\right)/2$ hard

4. Find the value of the integral $\int_0^1 \sqrt{1-x^2}\, dx$.

    A) $3\pi/4$ B) 3/4   C) 1/2   D) $\pi/4$

    E) 0   F) 1   G) $\pi/2$ H) 1/4

    Answer: $\pi/4$ medium

5. Find the value of the integral $\int_0^1 \sqrt{2x-x^2}\, dx$.

    A) $\pi/4$ B) 1/2   C) 1   D) $3\pi/4$

    E) 1/4   F) 3/4   G) 0   H) $\pi/2$

    Answer: $\pi/4$ hard

6. Find the value of the integral $\int_{1/2}^{1} \dfrac{1}{\sqrt{1-x^2}}\,dx$.

A) $\pi/2$ B) $\pi/6$ C) 0 D) $3\pi/2$

E) $3\pi/4$ F) π G) $2\pi/3$ H) $\pi/3$

Answer: $\pi/3$ medium

_____

7. Find the value of the integral $\int_{0}^{3/4} \dfrac{x}{\sqrt{x^2+1}}\,dx$.

A) 1 B) $3/4$ C) $\pi/2$ D) $1/4$

E) $\pi/4$ F) $3\pi/4$ G) 0 H) $1/2$

Answer: $1/4$ medium

_____

8. Find the value of the integral $\int_{0}^{3} \dfrac{dx}{\sqrt{9+x^2}}$.

A) $\sqrt{2}$ B) $3\sqrt{2}$ C) $1+\sqrt{2}$

D) $3+\sqrt{2}$ E) $\ln\left(1+\sqrt{2}\right)$ F) $\ln\left(3+\sqrt{2}\right)$

G) $\ln\left(\sqrt{2}-1\right)$ H) $\ln\left(3-\sqrt{2}\right)$

Answer: $\ln\left(1+\sqrt{2}\right)$ medium

_____

9. Find the value of the integral $\int_{0}^{2} \dfrac{x^3}{\sqrt{x^2+4}}\,dx$.

A) $\frac{8}{3}\left(2-\sqrt{2}\right)$ B) $\frac{10}{3}\left(2+\sqrt{2}\right)$ C) $4\left(3-2\sqrt{2}\right)$

D) $\frac{14}{3}\left(2\sqrt{2}-1\right)$ E) $\frac{16}{3}\left(5-2\sqrt{2}\right)$ F) $6\left(4-\sqrt{2}\right)$

G) $\frac{20}{3}\left(4+\sqrt{2}\right)$ H) $12\left(5+\sqrt{2}\right)$

Answer: $\frac{8}{3}\left(2-\sqrt{2}\right)$ medium

_____

10. Find the integral $\int \dfrac{dx}{x^2 \sqrt{x^2+4}}$.

A) $\sqrt{x^2+4} + C$

B) $x \sqrt{x^2+4} + C$

C) $-\dfrac{\sqrt{x^2+4}}{4x} + C$

D) $-\dfrac{\sqrt{x^2+4}}{x^2} + C$

E) $\dfrac{1}{\sqrt{x^2+4}} + C$

F) $\dfrac{x}{\sqrt{x^2+4}} + C$

G) $-\dfrac{1}{4x \sqrt{x^2+4}} + C$

H) $-\dfrac{x^2}{\sqrt{x^2+4}} + C$

Answer: $-\dfrac{\sqrt{x^2+4}}{4x} + C$ medium

11. Find the value of the integral $\int_0^3 x^2 \sqrt{9-x^2}\ dx$.

A) 1/8   B) 3/16   C) 5/32   D) 9/64

E) $27\pi/8$ F) $81\pi/16$ G) $243\pi/32$ H) $9\pi/64$

Answer: $81\pi/16$ medium

12. Find the integral $\int \dfrac{3x^2\ dx}{\sqrt{4-x^2}}$.

Answer: $6 \operatorname{Arcsin}\left(\dfrac{x}{2}\right) - \dfrac{3}{2} x\sqrt{4-x^2} + C$ medium

13. If an integral involves the quantity $\sqrt{x^2+1}$, a good substitution to consider is:

A) $x = \sin u$ B) $x = \tan u$ C) $x = \sec u$ D) $x = e^u$

E) $x = \ln u$

Answer: $x = \tan u$ easy

14. Find the integral $\int \dfrac{\sqrt{x^2-4}}{x}\ dx$.

Answer: $\sqrt{x^2-4} - 2 \operatorname{arc\ sec}\left(\dfrac{x}{2}\right) + C$ hard

15. Find the integral $\displaystyle\int_0^2 \frac{x^2}{(4+x^2)^2}\,dx$.

Answer: $\dfrac{\pi-2}{16}$ medium

---

16. Find the integral $\displaystyle\int \sqrt{1-x^2}\,dx$.

Answer: $\frac{1}{2}\sin^{-1}x + \frac{1}{2}x\sqrt{1-x^2} + C$ hard

---

17. Find the integral $\displaystyle\int \frac{1}{x\left(x^2+1\right)^{\frac{3}{2}}}\,dx$.

Answer: $\ln\left|\dfrac{\sqrt{x^2+1}}{x} - \dfrac{1}{x}\right| + \dfrac{1}{\sqrt{x^2+1}} + C$ hard

---

18. Find the integral $\displaystyle\int \sqrt{x^2+5}\,dx$.

Answer: $\dfrac{x\sqrt{x^2+5}}{2} + \dfrac{5}{2}\log\left(\sqrt{x^2+5}+x\right) + C$ hard

---

19. Find the integral $\displaystyle\int \frac{dx}{\sqrt{x^2+2x+2}}$.

Answer: $\ln\left|\sqrt{x^2+2x+2}+(x+1)\right| + C$ medium

---

20. Find the integral $\displaystyle\int \frac{dx}{2x\sqrt{9-x^2}}$.

Answer: $\frac{1}{6}\ln\left|\dfrac{3-\sqrt{9-x^2}}{x}\right| + C$ medium

---

Calculus, 2nd Edition
by James Stewart
Chapter 7, Section 4
Integration of Rational Functions by Partial Fractions

1. In the partial fraction decomposition $\dfrac{1}{x^2-1} = \dfrac{A}{x+1} + \dfrac{B}{x-1}$ find the value of A.

   A) -1　　　　B) $1/2$　　　　C) 2　　　　D) -2

   E) 1　　　　F) $3/4$　　　　G) $-1/2$　　　　H) $-3/4$

   Answer: $-1/2$　easy
   _____

2. In the partial fraction decomposition $\dfrac{x}{x^2-1} = \dfrac{A}{x+1} + \dfrac{B}{x-1}$ find the value of A.

   A) $-3/4$　　　　B) 1　　　　C) -1　　　　D) $1/2$

   E) 2　　　　F) -2　　　　G) $3/4$　　　　H) $-1/2$

   Answer: $1/2$　easy
   _____

3. Find the value of the integral $\displaystyle\int_0^{1/2} \dfrac{1}{x^2-1}\,dx$.

   A) $\frac{1}{2}\ln\frac{3}{4}$　　　　B) $\ln\frac{3}{4}$　　　　C) $\frac{1}{2}\ln\frac{4}{3}$　　　　D) $\frac{1}{2}\ln\frac{2}{3}$

   E) $\frac{1}{2}\ln 2$　　　　F) $\frac{1}{2}\ln\frac{1}{3}$　　　　G) $\frac{1}{2}\ln 3$　　　　H) $\ln\frac{1}{3}$

   Answer: $\frac{1}{2}\ln\frac{1}{3}$ medium
   _____

4. Find the value of the integral $\displaystyle\int_0^{1/2} \dfrac{x}{x^2-1}\,dx$.

   A) $\frac{1}{2}\ln\frac{3}{4}$　　　　B) $\ln\frac{3}{4}$　　　　C) $\ln 3$　　　　D) $\ln\frac{1}{3}$

   E) $\frac{1}{2}\ln\frac{1}{3}$　　　　F) $\frac{1}{2}\ln 3$　　　　G) $\frac{1}{2}\ln 2$　　　　H) $\frac{1}{2}\ln\frac{2}{3}$

   Answer: $\frac{1}{2}\ln\frac{3}{4}$　medium
   _____

5. Find the value of the integral $\int_0^{1/2} \frac{x^2}{x^2-1}\, dx$.

A) $\frac{1}{2}(1-\ln 3)$ B) $\ln 3$ C) $1-\ln\frac{3}{4}$ D) $1+\ln\frac{3}{4}$

E) $1+\ln 3$ F) $-\frac{1}{2}\ln 3$ G) $1-\ln 3$ H) $\frac{1}{2}\ln 3$

Answer: $\frac{1}{2}(1-\ln 3)$ medium

6. In the partial fraction decomposition of $\dfrac{1}{x^2(x+1)}$ find the numerator of the fraction whose denominator is x^2.

A) $1/2$ B) -1 C) -3 D) 2

E) 3 F) -2 G) $-1/2$ H) 1

Answer: 1 medium

7. Find the value of the integral $\int_0^1 \frac{1}{x^2-2x+2}\, dx$.

A) $\ln\frac{1}{2}$ B) 0 C) $3\pi/4$ D) $\pi/2$

E) $\ln\frac{3}{4}$ F) $\ln\frac{1}{4}$ G) $\pi/4$ H) 1

Answer: $\pi/4$ medium

8. Find the value of the integral $\int_0^1 \frac{2x^2\, dx}{(x+1)(x^2+1)}$.

A) $5\sqrt{5}-1$ B) $3\pi/2$ C) $\frac{\pi}{2}(3\sqrt{3}-1)$

D) $\frac{\ln 2}{3}$ E) $\frac{3}{2}\ln 2 - \frac{\pi}{4}$ F) $\ln 8 - \frac{\pi}{2}$

G) $\frac{\pi}{2}(\ln 3 - 1)$ H) $\ln 2 - \frac{\sqrt{3}}{2}$

Answer: $\frac{3}{2}\ln 2 - \frac{\pi}{4}$ hard

9. Find the value of the integral $\int_2^3 \frac{dx}{x(x-1)}$.

A) $3/2$ B) $4/3$ C) $\ln 2$ D) $\ln 3$

E) $\ln(3/2)$ F) $\ln(4/3)$ G) $\ln(2/3)$ H) $\frac{3}{2}\ln 2$

Answer: $\ln(4/3)$ medium

10. Find the value of the integral $\int_0^1 \frac{x}{x+1}\,dx$.

A) $1-\ln 2$ B) $1+\ln 2$ C) $\ln 2$ D) $-\ln 2$

E) $2-\ln 2$ F) $2+\ln 2$ G) $2+2\ln 2$ H) $2-2\ln 2$

Answer: $1-\ln 2$ medium

11. Find the value of the integral $\int_1^2 \frac{1}{x^3+x}\,dx$.

A) $3\ln 2-\ln 3$ B) $\ln 2-2\ln 3$ C) $\ln 2-\ln 5$

D) $3\ln 2-\ln 5$ E) $\dfrac{3\ln 2-\ln 3}{2}$ F) $\dfrac{\ln 2-2\ln 3}{2}$

G) $\dfrac{\ln 2-\ln 5}{2}$ H) $\dfrac{3\ln 2-\ln 5}{2}$

Answer: $\dfrac{3\ln 2-\ln 5}{2}$ medium

12. Find the value of the integral $\int_2^4 \frac{dx}{x^2-1}$.

A) $\ln 2$ B) $\ln 3$ C) $\ln 4$

D) $\ln 5$ E) $\ln 4-\ln 3$ F) $2\ln 5-\ln 4$

G) $\frac{1}{5}\ln 5-\frac{1}{3}\ln 3$ H) $\ln 3-\frac{1}{2}\ln 5$

Answer: $\ln 3-\frac{1}{2}\ln 5$ medium

13. Find the value of the integral $\int_0^1 \frac{x+1}{x^2+1}\,dx$.

A) π B) $1-2\ln 2$ C) $\frac{\pi}{2}-1$ D) $\dfrac{\ln 6}{2}$

E) $\dfrac{4-\pi}{2}$ F) $\dfrac{\pi+2\ln 2}{4}$ G) $\dfrac{4\ln 2-\pi}{6}$ H) $\dfrac{\pi-\ln 2}{8}$

Answer: $\dfrac{\pi+2\ln 2}{4}$ medium

14. Find the value of the integral $\int_3^4 \frac{dx}{(x-1)(x-2)}$.

A) $\ln 2$ B) $\ln 3$ C) $\ln 4$ D) $\ln 8$

E) $\ln(1/2)$ F) $\ln(2/3)$ G) $\ln(4/3)$ H) $\ln(8/3)$

Answer: $\ln(4/3)$ medium

15. Evaluate the integral $\int \dfrac{5x^2 + 26x + 29}{(x+2)(x+1)(x+3)}\,dx$.

Answer: $3\ln|x+2| + 4\ln|x+1| - 2\ln|x+3| + C$ hard!

16. Evaluate the integral $\int \dfrac{2x+1}{(x^2+1)(3x-1)}\,dx$.

Answer: $-\frac{1}{4}\ln(x^2+1) + \frac{1}{2}\,\text{Arctan } x + \frac{1}{2}\ln|3x-1| + C$ hard

17. Evaluate the integral $\int \dfrac{12 + 21x - 8x^2}{4x^2 - x^3}\,dx$.

Answer: $-\frac{3}{x} + 2\ln\left|x^3(4-x)\right| + C$ hard

18. Evaluate the integral $\int \dfrac{25}{x^4 + 2x^3 + 5x^2}\,dx$.

Answer: $\ln\!\left(\dfrac{x^2 + 2x + 5}{x^2}\right) - \frac{5}{x} - \frac{3}{2}\tan^{-1}\dfrac{x+1}{2} + C$ hard

19. Evaluate the integral $\int \dfrac{(2x+1)}{(x^2-1)(x^2+1)}\,dx$.

Answer: $\frac{3}{4}\ln|x+1| + \frac{1}{4}\ln|x-1| - \frac{1}{2}\ln\left|x^2+1\right| - \frac{1}{2}\,\text{Arctan } x + C$ hard

20. Evaluate the integral $\int \dfrac{1}{x^3 - x^2 + x - 1}\,dx$.

Answer: $-\frac{1}{2}\arctan x + \frac{1}{2}\ln(x-1) + C$ hard

21. Use the method of partial fractions to evaluate $\int_0^1 \dfrac{3x+4}{x^3-2x-4}\,dx$.

Answer: $\ln 10^{-1/2}$ hard

22. Evaluate the integral $\int \dfrac{2x^2+5x-4}{x(x+2)(x-1)}\,dx$.

Answer: $2\ln|x| - \ln|x+2| + \ln|x-1| + C$ hard

23. Evaluate the integral $\int \dfrac{6x^3+3x+1}{x^2(x^2+1)}\,dx$.

Answer: $3\ln|x| - \dfrac{1}{x} + \dfrac{3}{2}\ln(x^2+1) - \tan^{-1}x + C$ hard

24. Evaluate the integral $\int \dfrac{x^2+x+4}{(x+1)(x^2+3)}\,dx$.

Answer: $\ln|x+1| + \dfrac{1}{\sqrt{3}}\tan^{-1}\dfrac{x}{\sqrt{3}} + C$ hard

25. Evaluate the integral $\int \dfrac{dx}{x^3+2x^2+x}$.

Answer: $\ln\left|\dfrac{x}{x+1}\right| + \dfrac{1}{x+1} + C$ medium

26. Evaluate the integral $\int \dfrac{x}{(x-2)(x+3)}\,dx$.

Answer: $\dfrac{2}{5}\ln|x-2| + \dfrac{3}{5}\ln|x+3| + C$ medium

27. Evaluate the integral $\int \dfrac{2x}{(x^2+1)(x+1)^2}\,dx$.

Answer: $\dfrac{\tan^{-1}x + \dfrac{1}{x+1} + C \quad \text{medium}}{}$

Calculus, 2nd Edition
by James Stewart
Chapter 7, Section 5
Rationalizing Substitutions

1. Find the value of the integral $\int_0^1 \frac{x}{\sqrt{x+1}}\, dx$.

    A) $\frac{2}{3}(2+\sqrt{2})$ B) $\frac{4}{3}(\sqrt{2}+1)$ C) $\frac{2}{3}(\sqrt{2}+1)$ D) $\frac{2}{3}(2-\sqrt{2})$

    E) $\frac{4}{3}(2-\sqrt{2})$ F) $\frac{2}{3}(\sqrt{2}-1)$ G) $\frac{4}{3}(2+\sqrt{2})$ H) $\frac{4}{3}(\sqrt{2}-1)$

    Answer: $\frac{2}{3}(2-\sqrt{2})$ medium

2. Find the value of the integral $\int_0^1 \frac{1}{1+\sqrt{x}}\, dx$.

    A) $3(1-\ln 2)$ B) $\frac{2}{3}(1-\ln 2)$ C) $\frac{3}{2}(1-\ln 2)$

    D) $2(1+\ln 2)$ E) $2(1-\ln 2)$ F) $\frac{2}{3}(1+\ln 2)$

    G) $\frac{3}{2}(1+\ln 2)$ H) $3(1+\ln 2)$

    Answer: $2(1-\ln 2)$ medium

3. Find the value of the integral $\int_3^8 \frac{\sqrt{x+1}}{x}\, dx$.

    A) $\frac{1}{2}\ln\frac{4}{3}$ B) $\ln\frac{2}{3}$ C) $\frac{1}{2}\ln\frac{8}{3}$ D) $2-\ln\frac{3}{2}$

    E) $\frac{1}{2}\ln\frac{1}{3}$ F) $\ln\frac{8}{3}$ G) $\ln\frac{1}{3}$ H) $\ln\frac{4}{3}$

    Answer: $2-\ln\frac{3}{2}$ hard

4. Find the value of the integral $\int_3^8 \frac{1}{x\sqrt{x+1}}\, dx$.

    A) $\ln(3/2)$ B) $\ln(5/2)$ C) $\ln(11/5)$ D) $\ln(4/3)$

    E) $\ln(8/3)$ F) $\ln(11/3)$ G) $\ln(5/3)$ H) $\ln(11/2)$

    Answer: $\ln(3/2)$ hard

5. Find the value of the integral $\displaystyle\int_0^{\pi/2} \frac{1}{1+\sin x}\,dx$.

A) 3   B) 2   C) 4/3   D) 1

E) 5/3   F) 2/3   G) 8/3   H) 7/3

Answer: 1   hard

6. Find the value of the integral $\displaystyle\int_1^4 \frac{x+1}{\sqrt{x}}\,dx$.

A) 23/3   B) 8   C) 7   D) 20/3

E) 22/3   F) 19/3   G) 25/3   H) 6

Answer: 20/3   easy

7. Find the value of the integral $\displaystyle\int_2^6 \frac{\sqrt{x-2}}{x+2}\,dx$.

A) 1   B) 2   C) $4-\pi$ D) $\pi/2$

E) $2+\pi$ F) $3\ln 2$ G) $2\ln 2-1$ H) $4-\ln 2$

Answer: $4-\pi$ hard

8. Find the value of the integral $\displaystyle\int_0^3 \frac{x}{\sqrt{x+1}}\,dx$.

A) $\tan^{-1}3$ B) $\ln 2$ C) 8/3   D) 3π

E) 3   F) 3/2   G) $\pi/\sqrt{3}$ H) $\frac{1}{2}\ln 3$

Answer: 8/3   medium

9. Find the integral $\displaystyle\int \frac{\sqrt{x+2}-1}{\sqrt{x+2}+1}\,dx$.

Answer: $x+2-4\sqrt{x+2}+4\log\left|\sqrt{x+2}+1\right|+C$ hard

10. Find the integral $\displaystyle\int \frac{2+\sqrt[3]{x}}{\sqrt[3]{x}+\sqrt{x}}\,dx$.

Answer: $\frac{6}{5}x^{5/6}-\frac{3}{2}x^{2/3}+6x^{1/2}-9x^{1/3}+18x^{1/6}-18\ln\left|x^{1/6}+1\right|+C$ hard

11. The substitution $x = u^{12}$ changes $\int \frac{1}{\sqrt[3]{x} + \sqrt[4]{x}}\, dx$ into:

A) $\int \frac{1}{u^4 + u^3}\, du$ B) $\int \frac{12u^8}{u+1}\, du$ C) $\int \frac{u^9}{u+1}\, du$

D) $\int \frac{1}{u^2 + u}\, du$ E) $\int \frac{12u}{u^2 + 1}\, du$

Answer: $\int \frac{12u^8}{u+1}\, du$ easy
_____

12. Find the integral $\int \frac{\sqrt[6]{x}}{\sqrt{x} + \sqrt[3]{x}}\, dx$.

Answer: $\frac{3}{2}x^{2/3} - 2\sqrt{x} + 3\sqrt[3]{x} - 6\sqrt[6]{x} + 6\, \ln\left(\sqrt[6]{x} + 1\right) + C$ hard
_____

13. Find the value of the integral $\int_0^1 \frac{1}{1 + \sqrt[3]{x}}\, dx$.

Answer: $3\left(\ln 2 - \frac{1}{2}\right)$
_____

14. Find the integral $\int \frac{1}{x\sqrt{x+1}}\, dx$.

Answer: $\ln\left|\frac{\sqrt{x+1} - 1}{\sqrt{x+1} + 1}\right| + C$
_____

15. Find the integral $\int \frac{1}{x - \sqrt{x+2}}\, dx$.

Answer: $\frac{2}{3}\left[2\, \ln\left|\sqrt{x+2} - 2\right| + \ln\left(\sqrt{x+2} + 1\right)\right] + C$
_____

16. Evaluate the integral $\int_1^3 \frac{\sqrt{x-1}}{x+1}\, dx$.

Answer: $2\sqrt{2}(1 - \pi/4)$
_____

17. Find the integral $\displaystyle\int \frac{\sqrt[3]{x}+1}{\sqrt[3]{x}-1}\,dx$.

Answer: $x + 3x^{2/3} + 6\sqrt[3]{x} + 6\ln\left|\sqrt[3]{x}-1\right| + C$

18. Find the integral $\displaystyle\int \frac{x}{x^2 - \sqrt[3]{x^2}}\,dx$.

Answer: $\frac{3}{4}\ln\left|x^{4/3}-1\right| + C$

19. Find the integral $\displaystyle\int \frac{1}{\sqrt[3]{x}+\sqrt[4]{x}}\,dx$.

Answer: $\frac{3}{2}x^{2/3} - \frac{12}{7}x^{7/12} + 2\sqrt{x} - \frac{12}{5}x^{5/12} + 3\sqrt[3]{x} - 4\sqrt[4]{x} + 6\sqrt[6]{x} - 12\sqrt[12]{x} +$

$12\ln\left(\sqrt[12]{x}+1\right) + C$

20. Find the integral $\displaystyle\int \sqrt{\frac{x-1}{x}}\,dx$.

Answer: $\sqrt{x(x-1)} - \ln\sqrt{x} + \sqrt{(x-1)} + C$

21. Find the integral $\displaystyle\int \frac{\sin x}{\cos^2 x + \cos x - 6}\,dx$.

Answer: $\frac{1}{5}\ln\left|\frac{\cos x + 3}{\cos x - 2}\right| + C$

22. Find the integral $\displaystyle\int \frac{e^{3x}}{e^{2x}-1}\,dx$.

Answer: $e^x + \frac{1}{2}\ln\left|\frac{e^x-1}{e^x+1}\right| + C$

23. Find the integral $\int \dfrac{dx}{3 - 5 \sin x}$.

Answer: $\dfrac{1}{4} \ln \left| \dfrac{\tan(x/2) - 3}{3 \tan(x/2) - 1} \right| + C$

24. Find the integral $\int \dfrac{\sec x}{1 + \sin x} \, dx$.

Answer: $\dfrac{1}{2} \ln \left| \dfrac{1 + \tan(x/2)}{1 - \tan(x/2)} \right| + \dfrac{1}{1 + \tan(x/2)} - \dfrac{1}{\left[1 + \tan(x/2)\right]^2} + C$

Calculus, 2nd Edition
by James Stewart
Chapter 7, Section 6
Strategy for Integration

1. Find the value of the integral $\int_0^{\pi/4} \frac{1}{\cos^2 x} \, dx$.

   A) $\pi/3$ B) $1/2$ C) $\pi/2$ D) $\sqrt{3}/2$

   E) $2\pi/3$ F) 1 G) π H) $\sqrt{3}/4$

   Answer: 1 easy

2. Find the value of the integral $\int_0^{\pi/3} \frac{\sin x}{\cos^2 x} \, dx$.

   A) 1 B) $\pi/3$ C) $1/2$ D) π

   E) $\sqrt{3}/4$ F) $\pi/2$ G) $2\pi/3$ H) $\sqrt{3}/2$

   Answer: 1 easy

3. Find the value of the integral $\int_1^4 e^{-\sqrt{x}} \, dx$.

   A) $(2e+1)/(6e)$ B) $(2e+1)/(6e^2)$ C) $(2e-1)/(12e)$

   D) $(2e+1)/(12e^2)$ E) $(2e-1)/(12e^2)$ F) $(4e-6)/e^2$

   G) $(4e-6)/e$ H) $(2e+1)/(12e)$

   Answer: $(4e-6)/e^2$ medium

4. Find the value of the integral $\int_1^e \frac{(\ln x)^3}{x} \, dx$.

   A) 1 B) $e/2$ C) $1/3$ D) $1/2$

   E) e F) $e/3$ G) $1/4$ H) $e/4$

   Answer: $1/4$ medium

5. Find the value of the integral $\int_0^1 \frac{x}{x^2+1} \, dx$.

   A) $\frac{1}{2}\ln 2$ B) $3/4$ C) 1 D) $\ln 2$

   E) $1/4$ F) $3/2$ G) $\frac{1}{4}\ln 2$ H) $1/2$

   Answer: $\frac{1}{2}\ln 2$ easy

6. Find the value of the integral $\int_0^{\pi/4} \tan^2 x \, dx$.

A) $2+(\pi/2)$ B) $1-(\pi/2)$ C) $2+(\pi/4)$ D) $1+(\pi/4)$

E) $2-(\pi/4)$ F) $1-(\pi/4)$ G) $2-(\pi/2)$ H) $1+(\pi/2)$

Answer: $1-(\pi/4)$ medium

_____

7. Find the value of the integral $\int_0^1 \frac{x^2}{x+1} \, dx$.

A) $\ln 2 - \frac{1}{2}$ B) $\ln 2 - \frac{1}{4}$ C) $\ln 2 + \frac{1}{4}$ D) $\frac{1}{2} \ln 2 + \frac{1}{2}$

E) $\frac{1}{2} \ln 2 - \frac{1}{2}$ F) $\frac{1}{2} \ln 2 + \frac{1}{4}$ G) $\ln 2 + \frac{1}{2}$ H) $\frac{1}{2} \ln 2 - \frac{1}{4}$

Answer: $\ln 2 - \frac{1}{2}$ medium

_____

8. Find the value of the integral $\int_0^1 x^7 e^{-x^4} \, dx$.

A) $\frac{1}{4}(2+1/e)$ B) $\frac{1}{2}(2+2/e)$ C) $\frac{1}{2}(2-1/e)$

D) $\frac{1}{4}(1+1/e)$ E) $\frac{1}{4}(2-1/e)$ F) $\frac{1}{4}(1-2/e)$

G) $\frac{1}{2}(2+1/e)$ H) $\frac{1}{2}(2-2/e)$

Answer: $\frac{1}{4}(1-2/e)$ hard

_____

9. Find the value of the integral $\int_0^1 \ln(1+x^2) \, dx$.

A) $\ln 2$ B) $\pi/8$ C) $\pi/2 - 2 + \ln 2$

D) $2 - \ln 2$ E) $\pi/4 + \ln 2$ F) $\pi - 4$

G) $\pi - 2$ H) $\pi - \ln 2$

Answer: $\pi/2 - 2 + \ln 2$ hard

_____

10. Find the value of the integral $\int \frac{1 + \ln x}{x \ln x} \, dx$.

A) $\ln x + C$ B) $\ln \ln x + C$ C) $x + \ln x + C$

D) $\ln x + \ln \ln x + C$ E) $x/\ln x + C$ F) $\ln x/(x + \ln x) + C$

G) $x \ln x + C$ H) $x \ln \ln x + C$

Answer: $\ln x + \ln \ln x + C$ medium

_____

11. Find the value of the integral $\int \cos \sqrt{x} \, dx$.

A) $2 \sin \sqrt{x} + C$

B) $2 \sqrt{x} \cos \sqrt{x} + C$

C) $\sqrt{x} (\cos \sqrt{x} + \sin \sqrt{x}) + C$

D) $\dfrac{\cos \sqrt{x} + \sin \sqrt{x}}{\sqrt{x}} + C$

E) $2(\sqrt{x} \sin \sqrt{x} + \cos \sqrt{x}) + C$

F) $2(\sqrt{x} \cos \sqrt{x} + \sin \sqrt{x}) + C$

G) $\sqrt{x} \cos \sqrt{x} + \dfrac{\sin \sqrt{x}}{\sqrt{x}} + C$

H) $\sqrt{x} \sin \sqrt{x} + \dfrac{\cos \sqrt{x}}{\sqrt{x}} + C$

Answer: $2 (\sqrt{x} \sin \sqrt{x} + \cos \sqrt{x}) + C$ hard

12. Evaluate the following integral $\int \dfrac{4\sqrt{x}}{6+x} \, dx$.

Answer: $8\sqrt{x} - 8\sqrt{6} \tan^{-1}\left(\sqrt{\dfrac{x}{6}}\right) + C$ medium

13. Evaluate the following integral $\displaystyle\int_0^{1/4} \sec(\pi u) \tan(\pi u) \, du$.

Answer: $\frac{1}{\pi}(\sqrt{2} - 1)$ easy

14. Evaluate the following integral $\int x\left(1 + x^3\right)^2 \, dx$.

Answer: $\dfrac{x^2}{2} + \dfrac{2x^5}{5} + \dfrac{x^8}{8} + C$ medium

15. $\int \sin^5 x \, \cos x \, dx$ is:

A) $\frac{1}{12} \sin^6 x \, \cos^2 x + C$

B) $\frac{1}{6} \sin^6 x \, \cos x + C$

C) $-\sin^6 x + 5 \sin^4 x \, \cos^2 x + C$

D) $\frac{1}{6} \sin^6 x + C$

E) $\frac{1}{6} \sin^7 x + C$

Answer: $\frac{1}{6} \sin^6 x + C$ easy

16. Integrate $\int \dfrac{1}{\sqrt{9-4x^2}}\,dx$.

Answer: $\frac{1}{2}\sin^{-1}\left(\frac{2}{3}x\right)+C$ medium

17. Integrate $\int t \sin t \,dt$.

Answer: $-t\cos t + \sin t$ medium

18. Integrate $\int \dfrac{x^3+2x}{\sqrt{x^2-2}}\,dx$.

Answer: $\frac{1}{3}\left(x^2-2\right)^{3/2}+4\left(x^2-2\right)^{1/2}+C$ medium

19. Integrate $\int \dfrac{e^x+e^{-x}}{e^x-e^{-x}}\,dx$.

Answer: $\ln\left(e^x+e^{-x}\right)+C$ medium

20. Evaluate the following integral $\int \dfrac{x^3+4x^2+13x+3}{x^2+4x+13}\,dx$.

Answer: $\frac{1}{2}x^2+\tan^{-1}\left(\frac{x+2}{3}\right)+C$ hard

21. Give the integration technique most likely to work for the following integral. It is not necessary to do the integration.

$$\int x \sin x \,dx$$

Answer: Integration by parts. The two factors, x and $\sin x$, are dissimilar.   easy

22. Give the integration technique most likely to work for the following integral. It is not necessary to do the integration.

$$\int \frac{x+1}{x^2-4} \, dx$$

Answer: Partial fractions. The integrand is a rational function, and the denominator factors.

easy

23. Give the integration technique most likely to work for the following integral. It is not necessary to do the integration.

$$\int \frac{t}{\sqrt{t+2}} \, dt$$

Answer: Substitution, such as $u^2 = t+2$, to remove the radical. easy

24. Give the integration technique most likely to work for the following integral. It is not necessary to do the integration.

$$\int \frac{x^3 + x^2 + 1}{x-3} \, dx$$

Answer: Long division. The fraction is improper, since the numerator is of higher degree than

the denominator. After division, integrate the quotient and remainder using basic

formulas. medium

25. Give the integration technique most likely to work for the following integral. It is not necessary to do the integration.

$$\int \sqrt{1-x^2} \, dx$$

Answer: Trigonometric substitution $x = \sin \theta$ to remove the radical. $x = \cos \theta$ would work as

well. medium

26. Give the integration technique most likely to work for the following integral. It is not necessary to do the integration.

$$\int \ln(2x - 3)\, dx$$

Answer: Integration by parts with $u = \ln(2x - 3)$ and $dv = dx$ to change the form

of the integral. easy

27. Evaluate the integral $\int \dfrac{x^5}{\sqrt{1 + x^3}}\, dx$.

Answer: $\frac{2}{9}\left(1 + x^3\right)^{3/2} - \frac{2}{3}\left(1 + x^3\right)^{1/2} + C$ medium

Calculus, 2nd Edition
by James Stewart
Chapter 7, Section 7
Using Tables of Integrals

1. Find the value of the integral $\int_{-1}^{1} \sqrt{1-u^2} \, du$.

   A) 2
   B) 1
   C) $\pi/4$
   D) 1/4

   E) 1/2
   F) π
   G) $\pi/2$
   H) 2π

   Answer: $\pi/2$ easy

2. Find the value of the integral $\int_{-1}^{1} \sqrt{2-u^2} \, du$.

   A) $2-(\pi/2)$
   B) $2+(\pi/4)$
   C) $1-(\pi/4)$
   D) $1+(\pi/4)$

   E) $(\pi/2)-1$
   F) $2-(\pi/4)$
   G) $1+(\pi/2)$
   H) $2+(\pi/2)$

   Answer: $1+(\pi/2)$ medium

3. Find the value of the integral $\int_{\pi/4}^{\pi/2} \csc u \, du$.

   A) $-\frac{1}{2}\ln(\sqrt{2}-1)$
   B) $\ln(2-\sqrt{2})$
   C) $\frac{1}{2}\ln(\sqrt{2}-1)$

   D) $-\ln(2-\sqrt{2})$
   E) $\ln(\sqrt{2}-1)$
   F) $-\frac{1}{2}\ln 2 - (\sqrt{2})$

   G) $\frac{1}{2}\ln 2 - (\sqrt{2})$
   H) $-\ln(\sqrt{2}-1)$

   Answer: $-\ln(\sqrt{2}-1)$ medium

4. Find the value of the integral $\int_{1}^{2} \sqrt{u^2-1} \, du$.

   A) $\sqrt{3}+\frac{1}{2}\ln(4-\sqrt{3})$
   B) $\sqrt{3}-\frac{1}{2}\ln(2+\sqrt{3})$
   C) $\sqrt{3}-\frac{1}{2}\ln(4+\sqrt{3})$

   D) $\sqrt{3}+\frac{1}{2}\ln(2-\sqrt{3})$
   E) $\sqrt{3}-\frac{1}{2}\ln(4-\sqrt{3})$
   F) $\sqrt{3}+\frac{1}{2}\ln(4+\sqrt{3})$

   G) $\sqrt{3}+\frac{1}{2}\ln(2+\sqrt{3})$
   H) $\sqrt{3}-\frac{1}{2}\ln(2-\sqrt{3})$

   Answer: $\sqrt{3}-\frac{1}{2}\ln(2+\sqrt{3})$ medium

5. Find the value of the integral $\int_0^1 \tan^{-1} u\, du$.

A) $\frac{\pi}{4} - \ln 2$ B) $\frac{\pi}{2} + \frac{1}{2}\ln 2$ C) $\frac{\pi}{2} - \ln 2$ D) $\frac{\pi}{2} + \ln 2$

E) $\frac{\pi}{4} + \ln 2$ F) $\frac{\pi}{2} - \frac{1}{2}\ln 2$ G) $\frac{\pi}{4} - \frac{1}{2}\ln 2$ H) $\frac{\pi}{4} + \frac{1}{2}\ln 2$

Answer: $\frac{\pi}{4} - \frac{1}{2}\ln 2$ medium

6. Find the value of the integral $\int_0^{\pi/12} \sin^2 u\, du$.

A) $(\pi - 3)/18$ B) $(\pi - 3)/16$ C) $(\pi - 3)/24$

D) $(\pi - 3)/72$ E) $(\pi - 3)/36$ F) $(\pi - 3)/12$

G) $(\pi - 3)/48$ H) $(\pi - 3)/60$

Answer: $(\pi - 3)/24$ easy

7. Find the value of the integral $\int_1^e u^{10} \ln u\, du$.

A) $\left(10e^{10} + 1\right)/121$ B) $\left(10e^{10} - 1\right)/121$ C) $\left(11e^{10} + 1\right)/121$

D) $\left(11e^{10} - 1\right)/121$ E) $\left(10e^{11} + 1\right)/121$ F) $\left(11e^{11} + 1\right)/121$

G) $\left(11e^{11} - 1\right)/121$ H) $\left(10e^{11} - 1\right)/121$

Answer: $\left(10e^{11} + 1\right)/121$ medium

8. Evaluate the integral $\int \csc^3\left(\frac{x}{2}\right) dx$.

Answer: $-\csc\left(\frac{x}{2}\right)\cot\left(\frac{x}{2}\right) + \ln\left|\csc\left(\frac{x}{2}\right) - \cot\left(\frac{x}{2}\right)\right| + C$ medium

9. Evaluate the integral $\int \frac{\sqrt{4 - 3x^2}}{x}\, dx$.

Answer: $\sqrt{4 - 3x^2} - 2\ln\left|\frac{2 + \sqrt{4 - 3x^2}}{x}\right| + C$ medium

10. Evaluate the integral $\int \dfrac{\sin x \cos x}{\sqrt{1+\sin x}}\, dx$.

Answer: $-\frac{2}{3}(2-\sin x)\sqrt{1+\sin x}+C$ medium

11. Evaluate the integral $\int x^3 \sin^{-1}(x^2)\, dx$.

Answer: $\dfrac{2x^4-1}{8}\sin^{-1}(x^2)+\dfrac{x^2\sqrt{1-x^2}}{8}+C$ medium

12. Evaluate the integral $\int \dfrac{x^5}{x^2+\sqrt{2}}\, dx$.

Answer: $\frac{1}{4}x^4-\dfrac{1}{\sqrt{2}}x^2+\ln(x^2+\sqrt{2})+C$ medium

13. Evaluate the integral $\int \sin^6 2x\, dx$.

Answer: $-\frac{1}{12}\sin^5 2x \cos 2x - \frac{5}{48}\sin^3 2x \cos 2x - \frac{5}{64}\sin 4x + \frac{5}{16}x + C$ hard

14. Evaluate the integral $\int \dfrac{x}{\sqrt{x^2-4x}}\, dx$.

Answer: $\sqrt{x^2-4x}+2\ln\left|x-2+\sqrt{x^2-4x}\right|+C$ medium

15. Evaluate the integral $\int_0^{\infty} x^4 e^{-x}\, dx$.

Answer: 24   medium

Calculus, 2nd Edition
by James Stewart
Chapter 7, Section 8
Approximate Integration

1. Use the Trapezoidal Rule with $n = 1$ to approximate the integral $\int_0^1 \sqrt{x}\, dx$.

   A) 1/2   B) 9/16   C) 7/16   D) 1/4

   E) 3/8   F) 2/3   G) 1/3   H) 5/8

   Answer: 1/2 easy
   _____

2. Use Simpson's Rule with $n = 2$ to approximate the integral $\int_0^1 x^3\, dx$.

   A) 5/8   B) 1/3   C) 3/8   D) 2/3

   E) 7/16   F) 1/4   G) 9/16   H) 1/2

   Answer: 1/4 easy
   _____

3. Use the Trapezoidal Rule with $n = 2$ to approximate the integral $\int_0^1 x^3\, dx$.

   A) 5/16   B) 1/4   C) 1/2   D) 5/8

   E) 1/3   F) 7/16   G) 2/3   H) 3/8

   Answer: 5/16 medium
   _____

4. Use Simpson's Rule with $n = 4$ to approximate the integral $\int_1^5 \frac{1}{x}\, dx$.

   A) 73/45   B) 71/48   C) 61/35   D) 73/48

   E) 61/36   F) 59/36   G) 59/35   H) 71/45

   Answer: 73/45 medium
   _____

5. Use the Midpoint Rule with $n = 5$ to approximate $\int_1^2 \frac{1}{x}\, dx$.

   A) 0.6909   B) 0.6913   C) 0.6919   D) 0.6925

   E) 0.6928   F) 0.6932   G) 0.6937   H) 0.6945

   Answer: 0.6919 medium
   _____

6. Use the Midpoint Rule with $n = 4$ to approximate $\int_0^{\pi/4} \tan x\, dx$.

A) 0.2914   B) 0.3160   C) 0.3289   D) 0.3317

E) 0.3450   F) 0.3601   G) 0.3764   H) 0.3844

Answer: 0.3450   medium

7. Suppose using $n = 10$ to approximate the integral of a certain function by the Trapezoidal Rule results in an upper bound for the error equal to $1/10$. What will the upper bound become if we change to $n = 20$?

A) 1/10000   B) 1/100   C) 1/80   D) 1/160

E) 1/1000   F) 1/20   G) 1/40   H) 1/100000

Answer: 1/40   easy

8. Suppose using $n = 10$ to approximate the integral of a certain function by Simpson's Rule results in an upper bound for the error equal to $1/10$. What will the upper bound become if we change to $n = 20$?

A) 1/100   B) 1/10000   C) 1/40   D) 1/1000

E) 1/160   F) 1/20   G) 1/100000   H) 1/80

Answer: 1/160   easy

9. Evaluate $\int_0^2 \sqrt{x^3 + 2}\, dx$ using Simpson's Rule with $n = 6$.

Answer: approximately 3.86   medium

10. Use Simpson's rule with $n = 10$ to approximate $\int_0^1 \frac{1}{1+x^2}\, dx$.

Answer: approximately 0.7846   medium

11. Use the midpoint rule with 2 equal subdivisions to get an approximation for ln 5.

Answer: 1.5   easy

12. Use (a) the Trapezoidal Rule and (b) Simpson's Rule to approximate the given integral with the given value of n. (Round your answers to six decimal places.)

$$\int_0^2 e^x \, dx, \quad n = 8$$

Answer: (a) $T = 6.422298$ (b) $S = 6.389194$

13. Use (a) the Trapezoidal Rule and (b) Simpson's Rule to approximate the given integral with the given value of n. (Round your answers to six decimal places.)

$$\int_0^1 \frac{1}{1+x^2} \, dx, \quad n = 10$$

Answer: (a) $T = -.784981$ (b) $S = 0.785398$

14. Use (a) the Trapezoidal Rule and (b) Simpson's Rule to approximate the given integral with the given value of n. (Round your answers to six decimal places.)

$$\int_{-1}^2 xe^x \, dx, \quad n = 12$$

Answer: (a) $T = 8.240073$ (b) $S = 8.125593$

15. Use (a) the Trapezoidal Rule and (b) Simpson's Rule to approximate the given integral with the given value of n. (Round your answers to six decimal places.)

$$\int_0^1 \cos(x^2) \, dx, \quad n = 4$$

Answer: (a) $T = 0.895759$ (b) $S = 0.904501$

16. Use (a) the Trapezoidal Rule and (b) Simpson's Rule to approximate the given integral with the given value of n. (Round your answers to six decimal places.)

$$\int_0^{\pi/4} x \tan x \, dx, \quad n = 6$$

Answer: (a) $T = 0.189445$ (b) $S = 0.904501$

17. Use (a) the Trapezoidal Rule, (b) the Midpoint Rule, and (c) Simpson's Rule to approximate the given integral with the given value of n. (Round your answers to six decimal places.)

$$\int_0^2 \frac{1}{\sqrt{1+x^3}} \, dx, \quad n = 10$$

Answer: (a) $T = 1.401435$ (b) $M = 1.402556$ (c) $S = 1.402206$

18. Use (a) the Trapezoidal Rule and (b) Simpson's Rule to approximate the given integral with the given value of n. (Round your answers to six decimal places.)

$$\int_2^3 \frac{1}{\ln x} \, dx, \quad n = 10$$

Answer: (a) $T = 1.119061$ (b) $S = 1.118428$

19. Use (a) the Trapezoidal Rule and (b) Simpson's Rule to approximate the given integral with the given value of n. (Round your answers to six decimal places.)

$$\int_0^1 \ln(1 + e^x) \, dx, \quad n = 8$$

Answer: (a) $T = 0.984120$ (b) $S = 0.983819$

20. The widths (in meters) of a kidney-shaped swimming pool were measured at 2-m intervals as indicated in the figure. Use Simpson's Rule to estimate the area of the pool.

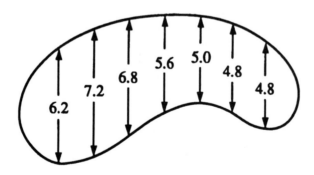

Answer: 84 m^2

Calculus, 2nd Edition
by James Stewart
Chapter 7, Section 9
Improper Integrals

1. Evaluate the improper integral $\int_1^\infty x^{-2}\, dx$.

   A) 1/4   B) 1   C) 2   D) 4

   E) 1/2   F) 3   G) 1/3   H) divergent

   Answer: 1 medium

2. Evaluate the improper integral $\int_0^1 x^{-2}\, dx$.

   A) 3   B) 1/4   C) 4   D) 2

   E) 1/3   F) 1/2   G) 1   H) divergent

   Answer: divergent easy

3. Evaluate the improper integral $\int_1^\infty x^{-1/2}\, dx$.

   A) 4   B) 3   C) 1/4   D) 1/3

   E) 1/2   F) 2   G) 1   H) divergent

   Answer: divergent easy

4. Evaluate the improper integral $\int_0^1 x^{-1/2}\, dx$.

   A) 1/3   B) 1/2   C) 1   D) 2

   E) 1/4   F) 3   G) 4   H) divergent

   Answer: 2 medium

5. Evaluate the improper integral $\int_1^\infty \frac{\ln x}{x}\, dx$.

   A) 2   B) $\frac{1}{2}\ln 2$ C) $\frac{1}{4}\ln 2$ D) 1

   E) $\ln 2$ F) 1/2   G) $2\ln 2$ H) divergent

   Answer: divergent medium

6. Evaluate the improper integral $\int_0^1 \frac{\ln x}{x}\, dx$.

   A) 1/2   B) 1   C) ln 2   D) $\frac{1}{4}$ ln 2

   E) $\frac{1}{2}$ ln 2   F) 2 ln 2   G) 2   H) divergent

   Answer: divergent medium
   _____

7. Evaluate the improper integral $\int_{-\infty}^{\infty} xe^{-x^2}\, dx$.

   A) e   B) $e^2 - 1$ C) 0   D) e^2
   E) e^{-1} F) e^{-2} G) 1   H) $1 - e^{-2}$

   Answer: 0 medium
   _____

8. Evaluate the improper integral $\int_{-1}^1 x^{-2}\, dx$.

   A) 1   B) 2   C) 4   D) 1/4
   E) 1/3   F) 3   G) 1/2   H) divergent

   Answer: divergent easy
   _____

9. Evaluate the improper integral $\int_0^2 \frac{dx}{(2x-3)}$.

   A) $-2/3$ B) $-1/3$ C) 0   D) 1/3
   E) 2/3   F) 1   G) 2   H) divergent

   Answer: divergent easy
   _____

10. Evaluate the improper integral $\int_{-\infty}^0 e^{3x}\, dx$.

    A) 3   B) 1/3   C) $-1/3$ D) 1
    E) -1 F) -3 G) 0   H) divergent

    Answer: 1/3 medium
    _____

11. Evaluate the improper integral $\int_0^{\infty} xe^{-x^2}\, dx$.

    A) 0   B) 1   C) e   D) e^{-1}
    E) $e - 1$ F) 1/2   G) 2   H) divergent

    Answer: 1/2 medium
    _____

12. Evaluate the improper integral $\int_1^\infty \frac{\ln x}{x^3}\,dx$.

A) 1/4   B) 1/3   C) 1/2   D) 1

E) ln 2   F) ln 3   G) ln 4   H) divergent

Answer: 1/4 medium

13. Evaluate the improper integral $\int_0^e \frac{dx}{x-1}$.

A) $\ln(e-1)$ B) e C) 0   D) 1

E) $-\ln(e-1)$ F) e^{-1} G) e^2 H) divergent

Answer: divergent easy

14. Evaluate the improper integral $\int_0^1 \frac{1}{3x-2}\,dx$.

A) 2   B) 3   C) $\frac{\ln 2}{3}$ D) $-\frac{\ln 2}{3}$

E) $\frac{\ln 4}{3}$ F) $-\frac{\ln 3}{2}$ G) $\frac{\ln 3}{2}$ H) divergent

Answer: divergent easy

15. Evaluate the improper integral $\int_1^\infty \frac{1}{(1+x)^4}\,dx$.

A) 1   B) 1/2   C) 1/4   D) 1/8

E) 1/16   F) 1/24   G) 1/32   H) divergent

Answer: 1/24 medium

16. Evaluate the improper integral $\int_1^\infty \frac{\ln x}{x^2}\,dx$.

A) 0   B) 1/4   C) 1/3   D) 1/2

E) 1   F) 2   G) 3   H) divergent

Answer: 1 medium

17. Evaluate the improper integral $\int_{-\infty}^\infty \frac{1}{1+x^2}\,dx$.

A) 1   B) 2   C) 3   D) π

E) 2π F) 3π G) $2/\pi$ H) $3/\pi$

Answer: π medium

18. Determine whether the improper integral is convergent or divergent. If it is convergent, evaluate it.

$$\int_{-\infty}^{2} \frac{1}{x^2 + 4}\, dx$$

Answer: $\frac{3\pi}{8}$ medium

19. Determine whether the improper integral is convergent or divergent. If it is convergent, evaluate it.

$$\int_{0}^{\infty} e^{-4x}\, dx$$

Answer: $\frac{1}{4}$ easy

20. Determine whether the improper integral is convergent or divergent. If it is convergent, evaluate it.

$$\int_{0}^{5} \left(\frac{1}{\sqrt{x}} + \frac{1}{\sqrt{5-x}} \right) dx$$

Answer: 6 medium

21. Determine whether the improper integral is convergent or divergent. If it is convergent, evaluate it.

$$\int_{3}^{\infty} \frac{1}{x^{3/2}}\, dx$$

Answer: $\frac{2}{\sqrt{3}}$ medium

22. Determine whether the improper integral is convergent or divergent. If it is convergent, evaluate it.

$$\int_{0}^{1} \frac{1}{(x-1)^2}\, dx$$

Answer: divergent medium

23. Determine whether the improper integral is convergent or divergent. If it is convergent, evaluate it.

$$\int_0^2 \frac{1}{(x-1)^2}\, dx$$

Answer: divergent medium

24. $\int_{-1}^3 \frac{1}{x}\, dx$ is:

A) $\frac{8}{9}$

B) $\ln 3$

C) $-1 + \ln 3$

D) $-\frac{10}{9}$

E) $\frac{12}{17}$

F) $\ln 9$

G) $1 + \ln 9$

H) a divergent improper integral

Answer: a divergent improper integral easy

25. Determine whether the improper integral is convergent or divergent. If it is convergent, evaluate it.

$$\int_3^\infty \frac{1}{x^2}\, dx$$

Answer: $\frac{1}{3}$ easy

26. Determine whether the improper integral is convergent or divergent. If it is convergent, evaluate it.

$$\int_{-1}^3 \frac{1}{x^2}\, dx$$

Answer: divergent easy

27. Determine whether the improper integral is convergent or divergent. If it is convergent, evaluate it.

$$\int_0^\infty \frac{x}{\left(x^2+5\right)^2}\, dx$$

Answer: $\frac{1}{10}$ medium

28. Determine whether the improper integral is convergent or divergent. If it is convergent, evaluate it.

$$\int_1^3 \frac{2}{(x-2)^{4/3}}\, dx$$

Answer: divergent medium

29. Determine whether the improper integral is convergent or divergent. If it is convergent, evaluate it.

$$\int_1^\infty x^{-5/4}\, dx$$

Answer: 4 medium

30. Determine whether the improper integral is convergent or divergent. If it is convergent, evaluate it.

$$\int_0^\infty \frac{8}{x^2+4}\, dx$$

Answer: 2π medium

31. Sketch $f(x) = \frac{1+\ln x}{x}$ and determine the area enclosed between the curve and the x-axis, as x approaches ∞.

Answer: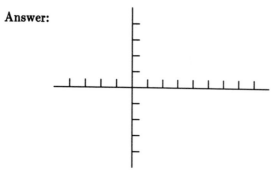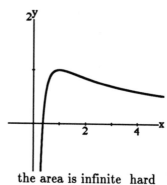

the area is infinite hard

Calculus, 2nd Edition
by James Stewart
Chapter 8, Section 1
Differential Equations

1. Solve the differential equation $y' = x^2$ subject to the initial condition $y(0) = 2$. From your solution find the value of $y(1)$.

    A) 7/2 B) 7/3 C) 10/3 D) 8/3

    E) 9/2 F) 5/2 G) 4 H) 3

    Answer: 7/3 easy
    _____

2. Solve the differential equation $y' = 5y(1000 - y)$ subject to the initial condition $y(0) = 500$. From your solution find the value of the limit $\lim\limits_{t \to \infty} y(t)$.

    A) 5000 B) 2500 C) 1000 D) 2000

    E) 200 F) 20000 G) 100 H) 500

    Answer: 1000 medium
    _____

3. Solve the differential equation $y' = y$ subject to the initial condition $y(0) = 0$. From your solution find the value of $y(e)$.

    A) e^e B) e C) $e^e - 1$ D) ln 2

    E) $e^e - e$ F) 0 G) e^2 H) 1

    Answer: 0 medium
    _____

4. Solve the differential equation $y' = e^{x-y}$ subject to the initial condition $y(0) = 2$. From your solution find the value of $y(1)$.

    A) $\ln\left(e^2 - e - 1\right)$ B) $\ln\left(e^2 - e + 1\right)$ C) $\ln\left(e^2 + e + 1\right)$

    D) $\ln\left(e^2 - e + 2\right)$ E) $\ln\left(e^2 + e - 1\right)$ F) $\ln\left(e^2 + e + 2\right)$

    G) $\ln\left(e^2 - e - 2\right)$ H) $\ln\left(e^2 + e - 2\right)$

    Answer: $\ln\left(e^2 + e - 1\right)$ hard
    _____

5. A tank contains 100 L of brine with 5 kg of dissolved salt. Pure water enters the tank at a rate of 10 L/min. The solution is kept thoroughly mixed and drains from the tank at the same rate. How much salt is in the tank after 6 minutes?

A) $5e^6$
B) $5e^{-6}$
C) $5e^{-0.6}$
D) $5e^{0.006}$

E) $5e^{-0.006}$
F) $5e^{-0.06}$
G) $5e^{0.06}$
H) $5e^{0.6}$

Answer: $5e^{-0.6}$ hard

6. A tank contains 100 L of pure water. Brine that contains 0.1 kg of salt per liter enters the tank at a rate of 10 L/min. The solution is kept thoroughly mixed and drains from the tank at the same rate. How much salt is in the tank after 6 minutes?

A) $100e^{-0.06}$
B) $100e^{-0.6}$
C) $10 - 10e^{-0.06}$

D) $10e^{-0.06}$
E) $10 - e^{-0.06}$
F) $10 - e^{-0.6}$

G) $10 - 10e^{-0.6}$
H) $10e^{-0.6}$

Answer: $10 - 10e^{-0.6}$ hard

7. Find the solution of the initial-value problem $y' = \frac{\ln x}{xy}$, $y(1) = 2$.

A) $y = \frac{1+x}{1+\ln x}$
B) $y = \frac{8x}{(1+x)^2}$
C) $y = 2 + 2\ln x$

D) $y = \sqrt{4 + (\ln x)^2}$
E) $y = x \ln x + 2x$
F) $y = x(1 + x^2)$

G) $y = x + \sqrt{1 + \ln x}$
H) $y = \sqrt{x}\,(1 + x)$

Answer: $\sqrt{4 + (\ln x)^2}$ medium

8. Solve the differential equation $y' = xy$.

Answer: $y = Ke^{x^2/2}$, where $K = \pm e^C$ is a constant   medium

9. Find the equation of the curve that passes through the point $(1, 1)$ and whose slope at (x, y) is y^2/x^3.

Answer: $y = \frac{2x^2}{1 + x^2}$ medium

10. Solve the differential equation: $\frac{dx}{dt} = 1 + t - x - tx$.

Answer: $x = 1 + Ae^{-(t^2/2 + t)}$ where $A = \pm e^C$ or 0 medium
_____

11. Solve the differential equation: $\frac{dy}{dx} = \frac{x + \sin x}{3y^2}$.

Answer: $y = \sqrt[3]{(x^2/2) - \cos x + C}$ easy
_____

12. Find a particular solution to the separable differential equation $x^2 y' = y^3$, $y(1) = 1$.

Answer: $y = \frac{x}{2 - x}$ medium
_____

13. Obtain a general solution of the following differential equation $\csc x \, dy + y^4 \, dx = 0$.

Answer: $\frac{1}{3y^2} = \cos x + C = 0$ medium
_____

14. Solve the differential equation $\frac{dz}{dt} = 2z^2 t \sqrt{1 + t^2}$, if $z = 1$ when $t = 0$.

Answer: $-\frac{1}{z} = \frac{2}{3}(1 + t^2)^{3/2} - \frac{5}{3}$ hard
_____

15. Find the general solution of the following differential equation $\frac{dy}{dx} = \frac{(3x + 1)^2 \sqrt{2 - y^2}}{y}$.

Answer: $\frac{1}{9}(3x + 1)^3 + \frac{1}{\sqrt{2 - y^2}} + C$ medium
_____

16. Solve the differential equation $e^{-y}y' + \cos x = 0$.

Answer: $y = -\ln|\sin x + C|$
_____

17. Solve the differential equation $y' = \dfrac{\ln x}{xy + xy^3}$.

Answer: $y^2 + 1 = \sqrt{2(\ln x)^2 + K}$

18. Solve $xy' = \sqrt{1 - y^2}$, $x > 0$, given $y(1) = 0$.

Answer: $y = \sin(\ln x)$

19. Solve $\dfrac{dy}{dx} = \dfrac{1+x}{xy}$ $x > 0$ given $y(1) = -4$.

Answer: $y^2 = 2 \ln x + 2x + 14$

20. Solve $\dfrac{dy}{dx} = e^{x-y}$ given $y(0) = 1$.

Answer: $y = \ln(e^x + e - 1)$

21. Solve $x\, dx + 2y\sqrt{x^2 + 1}\, dy = 0$ given $y(0) = 1$.

Answer: $y^2 = 2 - \sqrt{x^2 + 1}$

22. Solve $\dfrac{dy}{dx} = \dfrac{ty + 3t}{t^2 + 1}$, given $y(2) = 2$.

Answer: $y = -3 + \sqrt{5t^2 + 5}$

Calculus, 2nd Edition
by James Stewart
Chapter 8, Section 2
Arc Length

1.　　Find the arc length of the curve $3y = 4x$ from $(3, 4)$ to $(9, 12)$.

A) 13　　　　B) 10　　　　C) 8　　　　D) 14

E) 9　　　　F) 15　　　　G) 11　　　　H) 12

Answer: 10　easy
_____

2.　　Find the arc length of the curve $y^2 = x^3$ from $(0, 0)$ to $(1/4, 1/8)$.

A) 65/216　　　B) 29/108　　　C) 59/216　　　D) 37/108

E) 61/216　　　F) 31/108　　　G) 35/108　　　H) 71/216

Answer: 61/216　medium
_____

3.　　Find the arc length of the curve $y = \dfrac{x^3}{6} + \dfrac{1}{2x}$, $2 \le x \le 3$.

A) 15/4　　　B) 7/2　　　C) 19/4　　　D) 9/2

E) 5　　　　F) 4　　　　G) 17/4　　　H) 13/4

Answer: 13/4　medium
_____

4.　　Find the arc length of the curve $y = \sqrt{4 - x^2}$, $0 \le x \le 2$.

A) 2π　　　B) $3\pi/4$　　　C) π　　　D) $7\pi/4$

E) $3\pi/2$　　　F) $\pi/2$　　　G) $9\pi/4$　　　H) $5\pi/4$

Answer: π　medium
_____

5.　　Find the arc length of the curve $y = \ln(\cos x)$, $0 \le x \le \pi/3$.

A) $\ln\left(2 + \sqrt{2}\right)$　　　B) $\ln\left(1 + \sqrt{3}\right)$　　　C) $\ln\left(2 - \sqrt{2}\right)$　　　D) $\ln\left(2 - \sqrt{3}\right)$

E) $\ln\left(1 + \sqrt{2}\right)$　　　F) $\ln\left(2 + \sqrt{3}\right)$　　　G) $\ln\left(\sqrt{3} - 1\right)$　　　H) $\ln\left(\sqrt{2} - 1\right)$

Answer: $\ln\left(2 + \sqrt{3}\right)$　medium
_____

6. Find the arc length of the curve $y = x^2/2$, $0 \le x \le 1$.

A) $(\sqrt{2}+1-\ln(\sqrt{2}+1))/2$ B) $(\sqrt{2}+1-\ln(\sqrt{2}-1))/2$
C) $(\sqrt{2}-\ln(\sqrt{2}-1))/2$ D) $(\sqrt{2}-\ln(\sqrt{2}+1))/2$
E) $(\sqrt{2}+\ln(\sqrt{2}+1))/2$ F) $(\sqrt{2}+1+\ln(\sqrt{2}+1))/2$
G) $(\sqrt{2}+\ln(\sqrt{2}-1))/2$ H) $(\sqrt{2}+1+\ln(\sqrt{2}-1))/2$

Answer: $(\sqrt{2}+\ln(\sqrt{2}+1))/2$ hard

7. Find the length of the curve $y = x^3$, $0 \le x \le 1$.

A) $\sqrt{2}$ B) $1+(1/\sqrt{3})$ C) $\sqrt{3/2}$

D) $2-(1/\sqrt{2})$ E) $\int_0^1 (1+x^3)\,dx$ F) $\int_0^1 \sqrt{1+9x^4}\,dx$

G) $\int_0^1 \sqrt{4+x^2}\,dx$ H) $\int_0^1 \sqrt{9+x^3}\,dx$

Answer: $\int_0^1 \sqrt{1+9x^4}\,dx$ easy

8. Find the length of the curve $y = \frac{2}{3} x^{3/2}$, $0 \le x \le 3$.

A) $13/3$ B) $14/3$ C) 5 D) $16/3$
E) $17/3$ F) 6 G) $19/3$ H) $20/3$

Answer: $14/3$ medium

9. Find the length of the curve $y = \ln(\cos x)$, $0 \le x \le \pi/4$.

A) $\sqrt{2}$ B) $\sqrt{2}-1$ C) $1-(1/\sqrt{2})$ D) $(\sqrt{2}-1)/2$
E) $\ln(1/\sqrt{2})$ F) $\ln(1+\sqrt{2})$ G) $\ln(\sqrt{2})-1$ H) $\ln(\sqrt{2})-(1/2)$

Answer: $\ln(1+\sqrt{2})$ medium

10. Find the length of the curve $y = xe^x$, $0 \le x \le 1$.

A) e B) $\int_0^1 \sqrt{1+(x+1)^2\,e^{2x}}\,dx$ C) $\int_0^1 \sqrt{1+x^2\,e^{2x}}\,dx$

D) $\int_0^1 xe^x\,dx$ E) π F) $\int_0^1 (1+x^2\,e^{2x})\,dx$

G) $\int_0^1 (x+1)^2\,e^{2x}\,dx$ H) $\int_0^1 x^2\,e^{2x}\,dx$

Answer: $\int_0^1 \sqrt{1+(x+1)^2\,e^{2x}}\,dx$ medium

11. Find the length of the arc of the curve $9y^2 = 4(x-1)^3$ from $(1, 0)$ to $(5, 16/3)$.

Answer: $\dfrac{10\sqrt{5} - 2}{3}$ hard

12. Find the arc length of the curve $y = 2x^{3/2}$ between $x = 0$ and $x = 3$.

Answer: $\frac{2}{27}\left(28^{3/2} - 1\right)$ medium

13. Find the length of the curve $y = \frac{8}{3}x^{3/2}$ from the point $(1, 8/3)$ to the point $(4, 64/3)$.

Answer: $\frac{1}{24}\left(65^{3/2} - 17^{3/2}\right)$ medium

14. Set up, but do not evaluate, the equations and/or integrals to find the perimeter of the region bounded by the curve $y = x^2 - 2x$ and the x-axis.

Answer: $P = 2 + \displaystyle\int_0^2 \sqrt{1 + (2x - 2)^2}\, dx$ medium

15. Use Simpson's Rule with $n = 10$ to estimate the arc length of $y = x^4$, $0 \le x \le 2$.

Answer: 16.65 medium

16. Find the length of the arc of $y = 1 - x^{2/3}$ from $A(-8, -3)$ to $B(-1, 0)$.

Answer: $\frac{1}{27}\left(80\sqrt{10} - 13\sqrt{13}\right)$

17. Find the length of $y = \dfrac{x^3}{6} + \dfrac{1}{2x}$ for $1 \le x \le 2$.

Answer: $\dfrac{17}{12}$

18. Find the length of $y = \frac{x^2}{2} - \frac{\ln x}{4}$ for $2 \le x \le 4$.

Answer: $6 + \dfrac{\ln 2}{4}$

19. Find the length of $y = \ln(\sin x)$ for $\pi/6 \le x \le \pi/3$.

Answer: $\ln\left(1 = 2/\sqrt{3}\right)$

20. Set up, but do not evaluate, an integral for the length of $y = x^4 - x^2$, $-1 \le x \le 2$.

Answer: $L = \displaystyle\int_{-1}^{2} \sqrt{16x^6 - 16x^4 + 4x^2 + 1}\; dx$

21. Set up, but do not evaluate, an integral for the length of $y = \tan x$, $0 < x < \pi/4$.

Answer: $L = \displaystyle\int_{0}^{\pi/4} \sqrt{1 + \sec^4 x}\; dx$

22. The figure shows a telephone wire hanging between two poles at $x = -b$ and $x = b$. It takes the shape of a catenary with equation $y = a \cosh(x/a)$. Find the length of the wire.

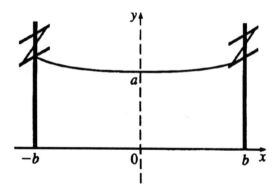

Answer: $2a \sinh(b/a)$

Calculus, 2nd Edition
by James Stewart
Chapter 8, Section 3
Area of a Surface of Revolution

1.   Find the area of the surface obtained by rotating the curve $y = 2x$, $0 \leq x \leq 1$, about the
     x-axis.

   A) $4\pi\sqrt{5}$ B) $\pi/\sqrt{5}$ C) $2\pi\sqrt{5}$ D) $4\pi/\sqrt{5}$

   E) $\pi\sqrt{5}$ F) $6\pi/\sqrt{5}$ G) $2\pi/\sqrt{5}$ H) $6\pi\sqrt{5}$

   Answer: $2\pi\sqrt{5}$ easy

2.   Find the area of the surface obtained by rotating the curve $y = 2x$, $0 \leq x \leq 1$, about the
     y-axis.

   A) $6\pi/\sqrt{5}$ B) $4\pi\sqrt{5}$ C) $4\pi/\sqrt{5}$ D) $\pi\sqrt{5}$

   E) $\pi/\sqrt{5}$ F) $6\pi\sqrt{5}$ G) $2\pi\sqrt{5}$ H) $2\pi/\sqrt{5}$

   Answer: $\pi\sqrt{5}$ easy

3.   Find the area of the surface obtained by rotating the curve $y = x^2/2$, $0 \leq x \leq 1$, about the
     y-axis.

   A) $(4\sqrt{2}-2)\pi/3$ B) $(2\sqrt{2}-2)\pi/3$ C) $(4\sqrt{2}+2)\pi/3$

   D) $(2\sqrt{2}+2)\pi/3$ E) $(2\sqrt{2}-1)\pi/3$ F) $(4\sqrt{2}+1)\pi/3$

   G) $(4\sqrt{2}-1)\pi/3$ H) $(2\sqrt{2}+1)\pi/3$

   Answer: $(4\sqrt{2}-2)\pi/3$ medium

4. Find the area of the surface obtained by rotating the curve $y = e^x$, $0 \le x \le \ln 2$, about the x-axis.

A) $\pi\left(2\sqrt{5} - \sqrt{2} + \ln\left(\dfrac{\sqrt{5}+2}{\sqrt{2}+1}\right)\right)$

B) $\pi\left(2\sqrt{5} + \sqrt{2} + \ln\left(\dfrac{2\sqrt{5}+2}{\sqrt{2}+1}\right)\right)$

C) $\pi\left(2\sqrt{5} + \sqrt{2} + \ln\left(\dfrac{\sqrt{5}+2}{\sqrt{2}+1}\right)\right)$

D) $\pi\left(2\sqrt{5} - \sqrt{2} - \ln\left(\dfrac{2\sqrt{5}+2}{\sqrt{2}+1}\right)\right)$

E) $\pi\left(2\sqrt{5} + \sqrt{2} - \ln\left(\dfrac{\sqrt{5}+2}{\sqrt{2}+1}\right)\right)$

F) $\pi\left(2\sqrt{5} - \sqrt{2} + \ln\left(\dfrac{2\sqrt{5}+2}{\sqrt{2}+1}\right)\right)$

G) $\pi\left(2\sqrt{5} - \sqrt{2} - \ln\left(\dfrac{2\sqrt{5}+2}{\sqrt{2}+1}\right)\right)$

H) $\pi\left(2\sqrt{5} - \sqrt{2} - \ln\left(\dfrac{\sqrt{5}+2}{\sqrt{2}+1}\right)\right)$

Answer: $\pi\left(2\sqrt{5} - \sqrt{2} + \ln\left(\dfrac{\sqrt{5}+2}{\sqrt{2}+1}\right)\right)$ hard

5. Find the area of the surface obtained by rotating the curve $y = \dfrac{x^3}{6} + \dfrac{1}{2x}$, $1 \le x \le 2$, about the y-axis.

A) $(15 + 2\ln 2)\pi/8$

B) $(15 + 4\ln 2)\pi/4$

C) $(15 + 4\ln 2)\pi/8$

D) $(31 + 2\ln 2)\pi/4$

E) $(31 + 4\ln 2)\pi/8$

F) $(31 + 2\ln 2)\pi/8$

G) $(31 + 4\ln 2)\pi/4$

H) $(15 + 2\ln 2)\pi/4$

Answer: $(15 + 4\ln 2)\pi/4$ medium

6. Find the area of the surface obtained by rotating the curve $y = \sqrt[3]{x}$, $0 \le x \le 1$, about the y-axis.

A) $26\pi/3$

B) $\pi(2\sqrt{2} - 1)/3$

C) $\pi(3\sqrt{3} - 1)/3$

D) $\pi(3\sqrt{3} - 1)/9$

E) $\pi(10\sqrt{10} - 1)/9$

F) $\pi(2\sqrt{2} - 1)/27$

G) $(3\sqrt{3} - 1)/27$

H) $\pi(10\sqrt{10} - 1)/27$

Answer: $\pi(10\sqrt{10} - 1)/27$ hard

7. The curve $y = \sqrt{x}$, $1 \le x \le 2$, is rotated about the x-axis. Find the area of the resulting surface.

A) $\pi(6\sqrt{6} - 1)/3$

B) $\pi(8 - 2\sqrt{2})/3$

C) $\pi(5\sqrt{5} - 2\sqrt{2})/3$

D) $\pi(3\sqrt{3} - 1)/3$

E) $\pi(7\sqrt{7} - 3\sqrt{3})/6$

F) $\pi(27 - 5\sqrt{5})/6$

G) $\pi(10\sqrt{10} - 16\sqrt{2})/6$

H) $\pi(17\sqrt{17} - 27)/6$

Answer: $\pi(27 - 5\sqrt{5})/6$ hard

8. The curve $y = x^3$, $0 \leq x \leq 1$, is rotated about the x-axis. Find the surface area of the resulting surface of revolution.

A) $26\pi/3$
B) $\pi(2\sqrt{2} - 1)/3$
C) $\pi(3\sqrt{3} - 1)/3$

D) $\pi(3\sqrt{3} - 1)/9$
E) $\pi(10\sqrt{10} - 1)/9$
F) $\pi(2\sqrt{2} - 1)/27$

G) $(3\sqrt{3} - 1)/27$
H) $\pi(10\sqrt{10} - 1)/27$

Answer: $\pi(10\sqrt{10} - 1)/27$ hard

9. Show that the surface obtained by rotating $y = \frac{1}{x}$ ($0 < x \leq 1$) about the y-axis has infinite surface area, but encloses a finite volume.

Answer: $S.A. = \int_0^1 2\pi x \sqrt{1 + \left(\frac{-1}{x^2}\right)^2}\, dx = \infty$; $V = \int_1^\infty \frac{\pi}{y^2}\, dy = \pi$ hard

10. Find the surface area when the graph of $f(x) = 3\sqrt{x}$, $0 \leq x \leq 2$, is rotated about the x-axis.

Answer: $\frac{\pi}{2}\left(17^{3/2} - 27\right)$ hard

11. Find the surface area generated when the quarter-circle $x^2 + y^2 = 4$ in the first octant is rotated around the y-axis.

Answer: 8π medium

12. Find the area of the surface generated by rotating the portion of the curve $y = \cosh x$, over the interval $0 \leq x \leq \ln 2$, about the x-axis.

Answer: $\pi\left(\frac{15}{16} + \ln 2\right)$ hard

13. A standard formula for a sphere of radius r is

$$\text{Surface area} = 4\pi r^2.$$

Regarding the sphere as a solid of revolution, prove this formula.

Answer: Rotate the circle of radius r with center at the origin to get the sphere.

An equation of the circle is $x^2 + y^2 = r^2$.

Then, $2x + 2y\dfrac{dy}{dx} = 0$, $\dfrac{dy}{dx} = -\dfrac{x}{y}$,

$$1 + \left(\frac{dy}{dx}\right)^2 = 1 + \frac{x^2}{y^2} = \frac{y^2 + x^2}{y^2} = \frac{r^2}{y^2}$$

Therefore,

$$\text{Surface area} = \int_b^a 2\pi y\sqrt{1 + \left(\frac{dy}{dx}\right)^2}\, dx = \ldots = 4\pi r^2 \quad \text{hard}$$

14. Find the surface area generated when the curve $y = \cosh x$, $0 \le x \le 1$, is rotated around the y-axis.

Answer: $2\pi\left(1 - \frac{1}{e}\right)$ hard

15. Find the area of the surface obtained by rotating $y^2 = 4x + 4$, $0 \le x \le 8$ about the x-axis.

Answer: $\frac{8\pi}{3}\left(10\sqrt{10} - 2\sqrt{2}\right)$

16. Find the area of the surface obtained by rotating $y = x^3$, $0 \le x \le 2$ about the x-axis.

Answer: $\frac{\pi}{27}\left(145\sqrt{145} - 1\right)$

17. Find the area of the surface obtained by rotating $y = \frac{x^2}{4} - \frac{\ln x}{2}$, $1 \le x \le 4$ about the x-axis.

Answer: $\pi\left[\frac{315}{16} - 8\ln 2 - (\ln 2)^2\right]$

18. Find the area of the surface obtained by rotating $y = \cos x$, $0 \le x \le \pi/3$ about the x-axis.

Answer: $\pi\left[\dfrac{\sqrt{21}}{4} + \ln\left(\dfrac{\sqrt{7}+\sqrt{3}}{2}\right)\right]$

19. Find the area of the surface obtained by rotating $x = \sqrt{2y - y^2}$, $0 \le y \le 1$ about the y-axis.

Answer: 2π

20. Find the area of the surface obtained by rotating $4x + 3y = 19$, $1 \le x \le 4$ about the y-axis.

Answer: 25π

21. Find the area of the surface obtained by rotating $x = a\,\cosh(y/a)$, $-a \le y \le a$ about the y-axis.

Answer: $2\pi a^2\left[1 + \tfrac{1}{2}\sinh 2\right]$

22. If the infinite curve $y = e^{-x}$, $x \ge 0$, is rotated about the x-axis, find the area of the resulting surface.

Answer: $\pi\left[\sqrt{2} + \ln\left(1 + \sqrt{2}\right)\right]$

Calculus, 2nd Edition
by James Stewart
Chapter 8, Section 4
Moments and Centers of Mass

1. Find the moment M_y of a system consisting of a mass $m_1 = 1$ at $(1, 0)$ and a mass $m_2 = 2$ at $(2, 0)$.

   A) 5   B) 3   C) 1   D) 6

   E) 10   F) 2   G) 4   H) 0

   Answer: 5 easy
   _____

2. Find the x-coordinate $\bar{x}$ at the center of mass of a system consisting of a mass $m_1 = 1$ at $(1, 0)$ and a mass $m_2 = 2$ at $(2, 0)$.

   A) 5/4   B) 11/6   C) 7/6   D) 1/2

   E) 7/4   F) 0   G) 5/3   H) 4/3

   Answer: 5/3 easy
   _____

3. Consider a flat plate of uniform density $\rho = 1$ bounded by the curves $y = x^2$ and $y = 1$. Find the moment M_x.

   A) 1.0   B) 0.6   C) 1.3   D) 0.7

   E) 1.1   F) 0.8   G) 1.2   H) 0.9

   Answer: 0.8 medium
   _____

4. Find the y-coordinate of the centroid of the region bounded by the curves $y = x^2$ and $y = 1$.

   A) 0.85   B) 0.70   C) 0.60   D) 0.50

   E) 0.75   F) 0.55   G) 0.80   H) 0.65

   Answer: 0.60 hard
   _____

5. Find the x-coordinate $\bar{x}$ of the centroid of the region bounded by the x-axis and the lines $y = x$ and $x = 2$.

A) 4/3   B) 7/6   C) 10/7   D) 5/3

E) 11/7   F) 11/6   G) 5/4   H) 7/4

Answer: 4/3 medium
_____

6. Find the volume obtained when a circle of radius 1 with center at $(1, 0)$ is rotated about the y-axis.

A) $3\pi^2$ B) $2\pi^2$ C) 3π D) 8π

E) $8\pi^2$ F) $4\pi^2$ G) 4π H) 2π

Answer: $2\pi^2$ medium
_____

7. Find the volume obtained when the square bounded by the lines $x = 0$, $x = 2$, $y = 1$, $y = -1$ is rotated about the y-axis.

A) 8π B) $3\pi^2$ C) $4\pi^2$ D) 3π

E) $8\pi^2$ F) 2π G) $2\pi^2$ H) 4π

Answer: 8π medium
_____

8. By Pappus' Theorem, the volume of the solid obtained by revolving the circle of radius 2 and center $(3, 7)$ about the x-axis is:

A) 28π B) 42π C) $56\pi^2$ D) $84\pi^2$

E) $19\pi^2$ F) $38\pi^3$ G) $76\pi^3$ H) $40\pi^{3/2}$

Answer: $56\pi^2$ medium
_____

9. Find the center of mass of the lamina of uniform density δ bounded by $y = 4 - x^2$ and the x-axis.

Answer: $\left(0, \frac{8}{5}\right)$ medium
_____

10. Determine the centroid of the region bounded by the equation $y^2 - 9x = 0$ in the first quadrant between $x = 1$ and $x = 4$.

Answer: $(2.65, 2.41)$ medium
_____

11. Find the center of mass of a homogeneous lamina of density k in the shape of the region bounded by $y = x^2$, $y = 0$, $x = 2$.

Answer: $\left(\frac{3}{2}, \frac{6}{5}\right)$ medium
_____

12. Find the centroid of the region bounded by the curves $y = 4x - x^2$ and $y = x$ in the xy-plane.

Answer: $\left(\frac{3}{2}, \frac{12}{5}\right)$ hard
_____

13. The masses m_i are located at the points P_i: $m_1 = 2$, $m_2 = 3$, $m_3 = 5$; $P_1(5, 1)$, $P_2(3, -2)$, $P_3(-2, 4)$. Find the moments M_x and M_y and the center of mass of the system.

Answer: $M_x = 0$, $M_y = 17$; $(\bar{x}, \bar{y}) = \left(\frac{17}{6}, 0\right)$
_____

14. Find the centroid of the region bounded by $y = 1 - x^2$ and $y - 0$.

Answer: $(\bar{x}, \bar{y}) = \left(0, \frac{2}{5}\right)$
_____

15. Find the centroid of the region bounded by $y = \sqrt{x}$, $y = 0$ and $x = 4$.

Answer: $(\bar{x}, \bar{y}) = \left(\frac{12}{5}, \frac{3}{4}\right)$
_____

16. Find the centroid of the region bounded by $y = \sin x$, $y = 0$, $x = 0$, and $x = \pi/2$.

Answer: $(\bar{x}, \bar{y}) = \left(1, \frac{\pi}{8}\right)$
_____

17. Find the centroid of the region bounded by $y = \ln x$, $y = 0$ and $x = e$.

Answer: $(\bar{x}, \bar{y}) = \left(\dfrac{e^2 + 1}{4}, \dfrac{e - 2}{2} \right)$ hard

18. Calculate the moments M_x and M_y and the center of mass of a lamina with $\rho = 5$ and shape given in the figure below.

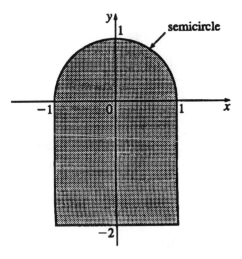

Answer: $M_x = \dfrac{2r^3}{3}$, $M_y = \dfrac{2r^3}{3}$, $(\bar{x}, \bar{y}) = \left(\dfrac{4r}{3\pi}, \dfrac{4r}{3\pi} \right)$

Calculus, 2nd Edition
by James Stewart
Chapter 8, Section 5
Hydrostatic Pressure and Force

1. An aquarium 1 foot high, 1 foot wide, and 2 feet long is filled with water. For simplicity, take the density of water to be 60 lb/ft^3. Find the hydrostatic pressure on the bottom of the aquarium in lb/ft^2.

    A) 30   B) 60   C) 14   D) 336
    E) 28   F) 120   G) 168   H) 240

    Answer: 60 easy
    _____

2. An aquarium 1 foot high, 1 foot wide, and 2 feet long is filled with water. For simplicity, take the density of water to be 60 lb/ft^3. Find the hydrostatic force in pounds on one of the 1 foot by 2 foot sides of the aquarium.

    A) 336   B) 30   C) 168   D) 240
    E) 60   F) 120   G) 28   H) 14

    Answer: 60 easy
    _____

3. A gate in an irrigation canal is in the form of a trapezoid 3 feet wide at the bottom, 5 feet wide at the top, with height equal to 2 feet. It is placed vertically in the canal, with the water extending to its top. For simplicity, take the density of water to be 60 lb/ft^3. Find the hydrostatic force in pounds on the gate.

    A) 360   B) 380   C) 440   D) 420
    E) 400   F) 460   G) 500   H) 480

    Answer: 440 medium
    _____

4. A right circular cylinder tank of height 1 foot and radius 1 foot is full of water. Taking the density of water to be a nice round 60 pounds per cubic foot, find the hydrostatic force in pounds on the side of the tank.

    A) 30π B) 60   C) 240π D) 240
    E) 30   F) 120π G) 60π H) 120

    Answer: 60π medium
    _____

5.   A right circular conical tank of height 1 foot and radius 1 foot at the top is full of water. Taking the density of water to be a nice round 60 pounds per cubic foot, find the hydrostatic force in pounds on the tank.

A) $60\sqrt{2}\pi$ B) $30\sqrt{2}$ C) $30\sqrt{2}\pi$ D) $20\sqrt{2}\pi$

E) $40\sqrt{2}\pi$ F) $60\sqrt{2}$ G) $40\sqrt{2}$ H) $20\sqrt{2}$

Answer: $20\sqrt{2}\pi$ hard

6.   A swimming pool 24 feet long and 15 feet wide has a bottom that is an inclined plane, the shallow end having a depth of 3 feet, and the deep end 10 feet. The pool is filled with water. For simplicity, take the density of water to be 60 lbs/ft^3. Find the hydrostatic force in pounds on the bottom of the pool.

A) 145750   B) 146250   C) 147000   D) 147250

E) 146500   F) 147500   G) 146750   H) 146000

Answer: 146250 hard

7.   A swimming pool 24 feet long and 15 feet wide has a bottom that is an inclined plane, the shallow end having a depth of 3 feet, and the deep end 10 feet. The pool is filled with water. For simplicity, take the density of water to be 60 lbs/ft^3. Find the hydrostatic force in pounds on one of the sides of the pool.

A) 33360   B) 40720   C) 39240   D) 36720

E) 40240   F) 40960   G) 40480   H) 39960

Answer: 33360 hard

8.   Find the total force on a submerged vertical plate in the form of an isosceles triangle with a base of 10 feet that lies 3 feet beneath the water surface and an altitude of 12 feet.

Answer: $F = 26,208$ medium

9.   Find the total hydraulic force on a dam in the shape of an equilateral triangle with a vertex down, if the side of the triangle is 100 feet and the water is even with the top.

Answer: $3906\frac{1}{4}$ tons medium

10. Assume water weighs 62.5 lb per cubic ft. Find the force due to water pressure on one side of a vertically submerged triangular plate having vertices at $(0, 0)$, $(1, 0)$, and $(1, 1)$ with the water surface at $y = 7$.

Answer: approximately 204.3 lb   hard

_____

11. A cylindrical barrel whose end has a diameter of 4 feet is submerged horizontally in seawater (64.3 lb/ft^3). Find the total force due to water pressure at one end if the center of the barrel is at a depth of 12 feet.

Answer: 3086.4π hard

_____

12. A flat plate of negligible thickness is in the shape of a right triangle with base 5' and height 10'. The plate is submerged in a tank of water. Find the force on the face of the plate under the following conditions. (Use 62.4 lb/ft^3 as the density of water.)

    The plate is submerged horizontally so that it rests flat on the bottom of the tank at a depth of 14'.

Answer: 1560 lb   easy

_____

13. A flat plate of negligible thickness is in the shape of a right triangle with base 5' and height 10'. The plate is submerged in a tank of water. Find the force on the face of the plate under the following conditions. (Use 62.4 lb/ft^3 as the density of water.)

    The plate is submerged vertically, base edge up and base at a depth of 3'.

Answer: 9880 lb   medium

_____

14. A swimming pool 5 m wide, 10 m long, and 3 m deep is filled with seawater of density 1030 kg/m^3 to a depth of 2.5 m. Find (a) the hydrostatic pressure at the bottom of the pool, (b) the hydrostatic force on the bottom, and (c) the hydrostatic force on one end of the pool.

Answer: (a) 25.2 kPa   (b) $1.26 \times 10^6 \text{ N}$ (c) $1.58 \times 10^5 \text{ N}$

15. A tank contains water. The end of the tank is vertical and has the shape below. Find the hydrostatic force against the end of the tank.

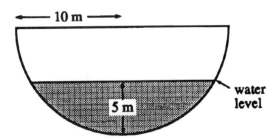

Answer: 1.23×10^6 N

16. A tank contains water. The end of the tank is vertical and has the shape below. Find the hydrostatic force against the end of the tank.

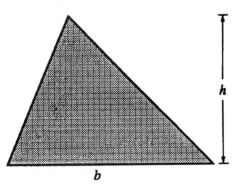

Answer: $1000gbh^2/3$ (metric units assumed)

17. A tank contains water. The end of the tank is vertical and has the shape below. Find the hydrostatic force against the end of the tank.

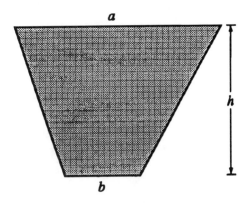

Answer: $\dfrac{500}{3}gh^2(a+2b)$ N

18. A vertical dam has a semicircular gate as shown in the figure. Find the hydrostatic force against the gate.

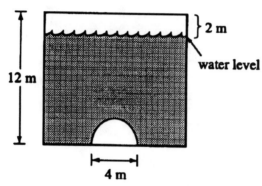

12 m

2 m

water level

4 m

Answer: 5.63×10^5 N

19. A dam is inclined an an angle of 30° from the vertical and has the shape of an isosceles trapezoid 100 ft wide at the top and 50 ft wide at the bottom and with a slant height of 70 ft. Find the hydrostatic force on the dam when it is full of water.

Answer: 7.71×10^6 lb

Calculus, 2nd Edition
by James Stewart
Chapter 8, Section 6
Applications to Economics and Biology

1.   The marginal revenue from selling x items is $90 - 0.02x$. The revenue from the sale of the first 100 items is $\$8800$. What is the revenue from the sale of the first 200 items?

Answer: $\$177,000$ easy

2.   The demand function for a certain commodity is $p = 5 - \frac{x}{10}$. Find the consumer's surplus when the sales level is 30. Illustrate by drawing the demand curve and identifying the consumer's surplus as an area.

Answer:

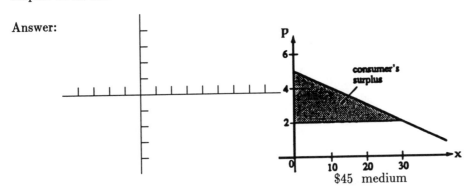

$\$45$ medium

3.   A trust fund pays $\$2000$ a year for 5 years, starting immediately. The interest rate is 12% per year compounded continuously. Find the present value of the trust fund.

Answer: $\$7519.81$ medium

4.   A baseball player signs a salary contract whereby he receives a sum that increases continuously and linearly from a starting salary of $\$1,000,000$ a year and reaches $\$3,000,000$ a year after 4 years. Thus his salary after t years (in millions of dollars) is $f(t) = 1 + \frac{1}{2}t$. Find the present value of the contract assuming an interest rate of 8% per year compounded continuously.

Answer: $\$4.72$ million   hard

5. An animal population is increasing at a rate of $200 + 50t$ per year(where t is measured in years). By how much does the animal population increase between the fourth and tenth years?

Answer: 3300 medium

6. Use Poiseuille's Law to calculate the rate of flow in a typical human artery where we can take $\eta = 0.027$, $R = 0.008$ cm, $l = 2$ cm, and $P = 4000$ dynes/cm^2.

Answer: see 8.6 #17 in the second edition medium

Calculus, 2nd Edition
by James Stewart
Chapter 9, Section 1
Curves Defined by Parametric Equations

1. Eliminate the parameter in the equations $x = t^2$, $y = t^4$.

    A) $y = x^2$ for $x \geq 0$ B) $y = \sqrt{x}$ for $x \geq 0$

    C) $y = 2x^2$ for $x \geq 0$ D) $y = \sqrt{2x}$ for $x \geq 0$

    E) $y = 2\sqrt{x}$ for $x \geq 0$ F) $y = x^2/2$ for $x \geq 0$

    G) $y = \sqrt{x}/2$ for $x \geq 0$ H) $y = \sqrt{x/2}$ for $x \geq 0$

    Answer: $y = x^2$ for $x \geq 0$ (easy)

2. Describe the curve defined by $x = \sin 2t$, $y = -\cos 2t$.

    A) circle   B) parabola   C) hyperbola

    D) cycloid   E) hypocycloid   F) involute

    G) trochoid   H) cardioid

    Answer: circle easy

3. Eliminate the parameter in the equations $x = \sin t$, $y = \sin^3 t$.

    A) $y = x^3$, $0 \leq x \leq 1$ B) $y = x^3$, $-1 \leq x \leq 1$

    C) $y = x^3$, $-1 \leq x \leq 0$ D) $y = \sqrt[3]{x}$, $0 \leq x \leq 1$

    E) $y = \sqrt[3]{x}$, $-1 \leq x \leq 1$ F) $y = \sqrt[3]{x}$, $-1 \leq x \leq 0$

    G) $y = x^{3/2}$, $0 \leq x \leq 1$ H) $y = x^{2/3}$, $-1 \leq x \leq 1$

    Answer: $y = x^3$, $-1 \leq x \leq 1$ medium

4. Describe the curve defined by $x = \sin t$, $y = \sin^2 t$.

    A) circle   B) semicircle

    C) quarter circle   D) parabola

    E) portion of parabola   F) hyperbola

    G) single branch of hyperbola   H) portion of branch of hyperbola

    Answer: portion of parabola medium

5. At how many places does the curve $x = \cos t$, $y = \sin 2t$ cross over the x-axis?

A) 5   B) 4   C) 7   D) 6

E) 0   F) 1   G) 3   H) 2

Answer: 3 hard

6. What kind of curve do the parametric equations $x = 2t - 3$, $y = 3t + 2$ describe?

A) a straight line   B) a circle   C) a cycloid

D) a parabola   E) an ellipse   F) a conchoid

G) a lemniscate   H) a spiral

Answer: a straight line easy

7. Sketch the curve represented by $x = \frac{t}{2}$, $y = 1 - t$.

Answer:

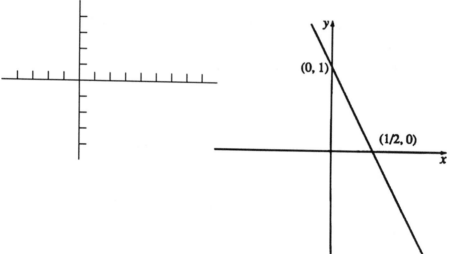

8. Sketch the curve represented by $x = \frac{1}{2} - \frac{1}{2} t^2$, $y = t^2$.

Answer:

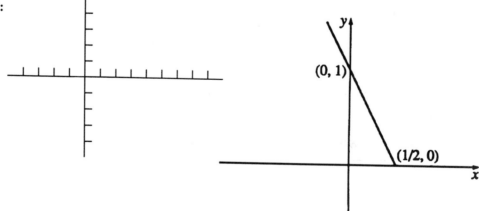

9.   Sketch the curve represented by $x = \frac{1}{2}\cos^2 t$, $y = \sin^2 t$.

Answer:

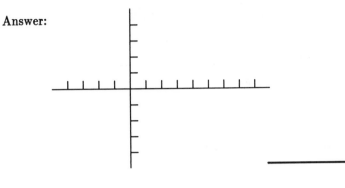

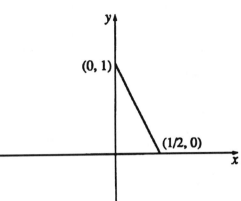

10.   Describe the motion of a particle with position (x, y) as t varies in the given interval.

$$x = 2 + \cos t, \quad y = 3 + \sin t, \quad 0 \le t \le 2\pi$$

Answer: the motion takes place on a unit circle centered at $(2,3)$. As t goes from 0 to 2π,

the particle makes one complete counterclockwise rotation around the circle, starting

and ending at $(3,3)$.   medium

11.   Sketch the curve represented by $x = t\cos t$, $y = t\sin t$.

Answer:

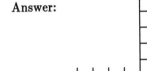

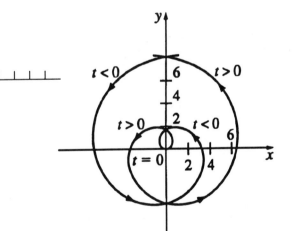

12. If a projectile is fired with an initial velocity of v_0 meters per second at an angle α above the horizontal, then its position after t seconds is given by the parametric equations

$$x = (v_0 \cos \alpha)\, t \qquad y = (v_0 \sin \alpha)\, t - \tfrac{1}{2} g t^2$$

   a) If a gun is fired with $\alpha = 30°$ and $v_0 = 500$ m/s, when will the bullet hit the ground?

   b) How far from the gun will it hit the ground?

   Answer: a) $t = 0$ and $t \doteq 51$ s; b) $x \doteq 22092$ m   hard

13. If a bullet is fired with an initial velocity of v_0 meters per second at an angle α above the horizontal, then its position after t seconds is given by the parametric equations

$$x = (v_0 \cos \alpha)\, t \qquad y = (v_0 \sin \alpha)\, t - \tfrac{1}{2} g t^2$$

   What is the maximum height reached by the bullet?

   Answer: 3189 m   medium

14. If a projectile is fired with an initial velocity of v_0 meters per second at an angle α above the horizontal, then its position after t seconds is given by the parametric equations

$$x = (v_0 \cos \alpha)\, t \qquad y = (v_0 \sin \alpha)\, t - \tfrac{1}{2} g t^2$$

   Show that the path is parabolic by eliminating the parameter.

   Answer: $y = (\tan \alpha)\, x - \dfrac{g}{2 v_0^2 \cos^2 \alpha} \cdot x^2$ which is the equation of a parabola.   hard

Calculus, 2nd Edition
by James Stewart
Chapter 9, Section 2
Tangents and Areas

1. Find the slope of the tangent to the curve $x = \sin t$, $y = \cos t$ when $t = \pi/3$.

   A) $-1/\sqrt{3}$　　B) $1/\sqrt{2}$　　C) $1/\sqrt{3}$　　D) 1

   E) -1　　F) $-\sqrt{3}$　　G) $\sqrt{3}$　　H) $-1/\sqrt{2}$

   Answer: $-\sqrt{3}$　medium

2. Find the slope of the tangent to the curve $x = t^3$, $y = t^4$ when $t = 3$.

   A) $9/4$　　B) $3/16$　　C) $4/9$　　D) $4/3$

   E) 3　　F) $3/4$　　G) $16/3$　　H) 4

   Answer: 4　easy

3. Find the slope of the tangent to the curve $x = \cos t$, $y = \cos^2 t$ when $t = 0$.

   A) $-1/2$　　B) -2　　C) 1　　D) $1/2$

   E) ∞　　F) -1　　G) 0　　H) 2

   Answer: 2　easy

4. At what value of t does the curve $x = t^2 - t$, $y = t^2 + t$ have a vertical tangent?

   A) $-1/3$　　B) $1/2$　　C) -2　　D) 1

   E) $-1/2$　　F) -3　　G) -1　　H) 2

   Answer: $1/2$　easy

5. At what value of t does the curve $x = t^2 - t$, $y = t^2 + t$ have a horizontal tangent?

   A) $-1/2$　　B) -2　　C) $1/2$　　D) 2

   E) $-1/3$　　F) 1　　G) -3　　H) -1

   Answer: $-1/2$　easy

6. Given $x = e^t$, $y = \sin t$, find the value of d^2y/dx^2 when $t = 0$.

A) $-1/\sqrt{2}$ B) -2 C) $1/\sqrt{2}$ D) $1/2$

E) -1 F) 2 G) 1 H) $-1/2$

Answer: -1 medium

7. Find the slope of the tangent to the curve with parametric equations $x = 2\ln t$, $y = t\, e^t$ at the point where $t = 1$.

A) 1 B) 2 C) 3 D) 4

E) e F) $2e$ G) $3e$ H) $4e$

Answer: e medium

8. Find the slope of the tangent to the curve with parametric equations $x = t + t^2$, $y = t + e^t$ at the point $(0, 1)$.

A) -3 B) -2 C) -1 D) 0

E) 1 F) 2 G) 3 H) 4

Answer: 2 medium

9. Find the slope of the tangent to the curve with parametric equations $x = \ln t$, $y = t^3$ at the point where $t = 1$.

A) 1 B) 2 C) 3 D) $1/2$

E) 4 F) 5 G) 6 H) 7

Answer: 3 medium

10. Find the slope of the tangent to the astroid with parametric equations $x = \cos^3 t$, $y = \sin^3 t$ at the point where $t = \pi/4$.

A) 1 B) -1 C) 3 D) $1/3$

E) -3 F) $-1/3$ G) $\pi/4$ H) 0

Answer: -1 medium

11. Find the slope of the tangent line to the curve with parametric equations $x = 1 + \ln t$, $y = t^2 - 3t$ at the point where $t = 3$.

A) 1   B) 3   C) 5   D) 7

E) 9   F) 11   G) 13   H) 15

Answer: 9 medium

12. Consider the curve given by $x = t^2 + 3$ and $y = 2t^3 - t$. Find $\dfrac{dy}{dx}$ at the point corresponding to $t = 2$.

Answer: $\dfrac{23}{4}$ easy

13. Consider the curve given by $x = t^2 + 3$ and $y = 2t^3 - t$. Find $\dfrac{d^2y}{dx^2}$ at the point corresponding to $t = 2$.

Answer: $\dfrac{25}{32}$ medium

14. Find an equation in x and y for the tangent line to the curve $x = e^t$, $y = e^{-t}$ at the point $\left(\dfrac{1}{3}, 3\right)$.

Answer: $y = -9x + 6$ medium

15. Find $\dfrac{dy}{dx}$: $x = t^4 - t^2 + t$, $y = \sqrt[3]{t}$.

Answer: $\dfrac{1}{3t^{2/3}\left(4t^3 - 2t + 1\right)}$ hard

16. Find $\dfrac{d^2y}{dx^2}$: $x = t^4 - t^2 + t$, $y = \sqrt[3]{t}$.

Answer: $\dfrac{-44t^3 + 10t - 2}{9t^{5/3}\left(4t^3 - 2t + 1\right)^3}$ hard

17. At what point does the curve $x = 1 - 2\cos^2 t$, $y = (\tan t)\left(1 - 2\cos^2 t\right)$ cross itself? Find the equations of both tangents at that point.

Answer: $(0, 0)$; $y = x$ and $y = -x$ hard

Calculus, 2nd Edition
by James Stewart
Chapter 9, Section 3
Arc Length and Surface Area

1. Find the length of the curve $x = 2t$, $y = 3t$, $0 \le t \le 1$.

    A) $\sqrt{26}$ B) $\sqrt{60}$ C) $\sqrt{14}$ D) $\sqrt{28}$

    E) $\sqrt{13}$ F) $\sqrt{15}$ G) $\sqrt{30}$ H) $\sqrt{52}$

    Answer: $\sqrt{13}$ easy
    _____

2. Find the length of the curve $x = 2t^2 - 1$, $y = 4t^2 + 3$, $0 \le x \le 2$.

    A) $4\sqrt{5}$ B) $8\sqrt{5}$ C) $8\sqrt{3}$ D) $2\sqrt{3}$

    E) $2\sqrt{5}$ F) $4\sqrt{3}$ G) $16\sqrt{3}$ H) $16\sqrt{5}$

    Answer: $8\sqrt{5}$ medium
    _____

3. Find the length of the curve $x = t^2$, $y = t^3$, $0 \le t \le \frac{1}{2}$.

    A) 59/128   B) 61/128   C) 61/108   D) 61/256

    E) 61/216   F) 59/256   G) 59/216   H) 59/108

    Answer: 61/216 hard
    _____

4. Find the length of the curve $x = \cos^2 t$, $y = \sin^2 t$, $0 \le t \le \pi$.

    A) $\pi/2$ B) $4\sqrt{2}$ C) $\pi/4$ D) $1/\sqrt{2}$

    E) $2\sqrt{2}$ F) π G) 2π H) $\sqrt{2}$

    Answer: $2\sqrt{2}$ medium
    _____

5. Find the length of the curve $x = \ln \cos t$, $y = t$, $0 \le t \le \pi/4$.

    A) $\ln\left(2\sqrt{2}+1\right)$ B) $\ln\left(\sqrt{2}+2\right)$ C) $\ln\left(2\sqrt{2}+2\right)$

    D) $\ln\left(\sqrt{2}+1\right)$ E) $\ln\left(2\sqrt{3}+1\right)$ F) $\ln\left(\sqrt{3}+2\right)$

    G) $\ln\left(2\sqrt{3}+2\right)$ H) $\ln\left(\sqrt{3}+1\right)$

    Answer: $\ln\left(\sqrt{2}+1\right)$ medium
    _____

6. Find the area of the surface obtained by rotating the curve $x = \sin t$, $y = \sin^2 t$ about the y-axis.

A) $4\pi\left(\sqrt{125/16} + 1\right)/3$

B) $\pi\left(\sqrt{125} + 1\right)/3$

C) $4\pi\left(\sqrt{125/64} + 1\right)/3$

D) $\pi\left(\sqrt{125} - 1\right)/3$

E) $2\pi\left(\sqrt{125/64} - 1\right)/3$

F) $4\pi\left(\sqrt{125/16} - 1\right)/3$

G) $2\pi\left(\sqrt{125/16} - 1\right)/3$

H) $2\pi\left(\sqrt{125/64} + 1\right)/3$

Answer: $\pi\left(\sqrt{125} - 1\right)/3$ hard

---

7. Give an integral representing the length of the parametric curve $x = t^3$, $y = t^4$, $0 \le t \le 1$.

A) $\displaystyle\int_0^1 \left(t^3 + t^4\right) dt$

B) $\displaystyle\int_0^1 \sqrt{t^3 + t^4}\, dt$

C) $\displaystyle\int_0^1 \sqrt{1 + 3t^2}\, dt$

D) $\displaystyle\int_0^1 \sqrt{t^2 + 4t^3}\, dt$

E) $\displaystyle\int_0^1 \sqrt{9t^4 + 16t^6}\, dt$

F) $\displaystyle\int_0^1 \sqrt{4t^4 + 9t^6}\, dt$

G) $\displaystyle\int_0^1 \sqrt{t^5 + t^7}\, dt$

H) $\displaystyle\int_0^1 \sqrt{8t^6 + 6t^8}\, dt$

Answer: $\displaystyle\int_0^1 \sqrt{9t^4 + 16t^6}\, dt$ easy

---

8. Find the length of the curve with parametric equations $x = t^3$, $y = t^2$, $0 \le t \le 1$.

A) $\dfrac{13\sqrt{13} - 1}{81}$

B) $\dfrac{13\sqrt{13} - 8}{27}$

C) $\dfrac{2\left(6\sqrt{6} - 1\right)}{27}$

D) $\dfrac{5\left(6\sqrt{6} - 1\right)}{18}$

E) $\dfrac{2\sqrt{2} - 1}{18}$

F) $\dfrac{2\left(5\sqrt{5} - 2\sqrt{2}\right)}{9}$

G) $\dfrac{4\left(10\sqrt{10} - 1\right)}{9}$

H) $\dfrac{15\sqrt{15} - 27}{2}$

Answer: $\dfrac{13\sqrt{13} - 8}{27}$ hard

---

9. Find the total length of the astroid $x = a\cos^3\theta$, $y = a\sin^3\theta$.

A) a

B) $2a$

C) $3a$

D) $4a$

E) $5a$

F) $6a$

G) $7a$

H) $8a$

Answer: $6a$ hard

---

10. Find the length of the curve with parametric equations $x = \cos t + t \sin t$, $y = \sin t - t \cos t$, $0 \le t \le \pi$.

A) π 　　　　　　 B) π^2 　　　　　　 C) $\pi/2$ 　　　　　　 D) $\pi^2/2$

E) $\pi/4$ 　　　　　　 F) $\pi^2/4$ 　　　　　　 G) $\pi/8$ 　　　　　　 H) $\pi^2/8$

Answer: $\pi^2/2$ medium

11. Find the length of the curve with parametric equations $x = \cos t$, $y = \sin t$, $0 \le t \le \pi/4$.

A) 1 　　　　　　 B) 2 　　　　　　 C) 3 　　　　　　 D) $\pi/8$

E) $\pi/4$ 　　　　　　 F) $\pi/2$ 　　　　　　 G) $3\pi/4$ 　　　　　　 H) π

Answer: $\pi/4$ easy

12. Write the definite integral representing the length of the curve with parametric equations $x = f(t)$, $y = g(t)$, $a \le t \le b$.

A) $\displaystyle\int_a^b 2\pi t\, f(t)\, dt$ 　　　　　　 B) $\displaystyle\int_a^b \pi\big[f(t)\big]^2 dt$

C) $\displaystyle\int_a^b \left(1 + [f'(t)]^2\right) dt$ 　　　　　　 D) $\displaystyle\int_a^b \big[1 + f'(t)\big]^2 dt$

E) $\displaystyle\int_a^b \sqrt{1 + \big[f'(t)\big]^2}\, dt$ 　　　　　　 F) $\displaystyle\int_a^b \sqrt{1 + \big[g'(t)\big]^2}\, dt$

G) $\displaystyle\int_a^b \sqrt{\big[f'(t)\big]^2 + \big[g'(t)\big]^2}\, dt$ 　　　　　　 H) $\displaystyle\int_a^b \big[f'(t) + g'(t)\big]^2 dt$

Answer: $\displaystyle\int_a^b \sqrt{\big[f'(t)\big]^2 + \big[g'(t)\big]^2}\, dt$ easy

13. What is the maximum curvature on the curve given by $y = x^2$?

Answer: $\kappa(0) = 2$ hard

14. Find the distance traveled by a particle with position (x, y) as t varies in the given time interval. Compare with the length of the curve.

$$x = \cos^2 t, \quad y = \cos t, \quad 0 \le t \le 4\pi$$

Answer: distance $= 4\sqrt{5} + 2\ln(\sqrt{5} + 2)$; $L = \sqrt{5} + \tfrac{1}{2}\ln(\sqrt{5} + 2)$ hard

15. Find the area of the surface obtained by rotating the given curve about the x-axis.

$$x = t^3, \ y = t^2, 0 \le t \le 1$$

Answer: $2\pi\left(247\sqrt{13} + 64\right)/1215$ medium

16. Find the area of the surface obtained by rotating the given curve about the x-axis.

$$x = a \cos^3\theta, \ y = a \sin^3\theta, 0 \le \theta \le \frac{\pi}{2}$$

Answer: $\frac{6\pi a^2}{5}$ medium

17. Find the surface area generated by rotating the given curve about the y-axis.

$$x = e^t - t, \ y = 4e^{t/2}, 0 \le t \le 1$$

Answer: $\pi\left(e^2 + 2e - 6\right)$ medium

18. Find the length of the curve represented by $x = \arcsin\frac{t}{2}$ and $y = \ln\sqrt{4 - t^2}, 0 \le t \le 1$.

Answer: $\ln\sqrt{3}$ medium

19. Find the circumference of the circle: $x = 2\cos t, \ y = 2\sin t$.

Answer: 4π medium

20. If $x = \cos 2t, \ y = \sin^2 t$, and (x, y) represents the position of a particle, find the distance the particle travels as t moves from 0 to $\frac{\pi}{2}$.

Answer: $\sqrt{5}$ hard

21. The involute of a circle of radius 1 is given parametrically by:

$$x = \cos t + t \sin t$$
$$y = \sin t - t \cos t$$

Find the length of the portion of the involute which is traced out as t increases from 0 to π.

Answer: $\frac{\pi^2}{2}$ medium

22. Compute the length of the curve given parametrically by $x = \frac{1}{3}t^3$ and $y = \frac{1}{2}t^2$ for $0 \le t \le 2$.

Answer: $\dfrac{5\sqrt{5}-1}{3}$ medium

23. Find the length of the curve given parametrically by $x = \frac{t^2}{2} + 7$ and $y = \frac{1}{3}(2t+1)^{3/2}$ for $2 \le t \le 6$.

Answer: 20 hard

24. An arc is described by the parametric equations $x = 3t^3$ and $y = 2t^2$. Sketch the arc and calculate its length from (a) $t = 0$ to $t = 2$, and (b) $t = -3$ to $t = 0$.

Answer:

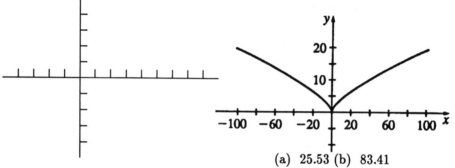

(a) 25.53 (b) 83.41

25. The equation of a curve in parametric form is

$$x = 5\cos 3t$$
$$y = 4\sin 3t$$

Find the arc length of the curve from $t = 0$ to $t = \frac{\pi}{8}$.

Answer: $\dfrac{3\pi}{2}$ medium

26. Find the length of the plane curve given parametrically by $x = t$ and $y = t^{2/3}$ between $t = 0$ and $t = 8$.

Answer: $\frac{8}{27}\left(10\sqrt{10} - 1\right)$ medium

27. A curve is written parametrically as $x = t^3$, $y = t^{9/2}$. Find the length of the curve from $t = 0$ to $t = 1$.

Answer: $\frac{8}{27}\left(\frac{13}{8}\sqrt{13} - 1\right)$ medium

28. Find the arc length of the curve defined by

$$x = t^3 + 1$$
$$y = 3t^2 + 2$$

for $0 \leq t \leq 1$.

Answer: $5\sqrt{5} - 8$ medium

29. The position of a particle $P(x, y)$ at time t is given by $x = \frac{1}{3}(2t + 3)^{3/2}$ and $y = \frac{t^2}{2} + t$. Find the arc length of the path the particle would travel from $t = 0$ to $t = 3$.

Answer: $\frac{21}{2}$ hard

Calculus, 2nd Edition
by James Stewart
Chapter 9, Section 4
Polar Coordinates

1. Convert the polar coordinates $(3, 5\pi)$ to Cartesian coordinates.

   A) $\left(-\sqrt{3}/2, -\sqrt{3}/2\right)$ B) $(3, 0)$ C) $(0, 3)$

   D) $\left(\sqrt{3}/2, \sqrt{3}/2\right)$ E) $(-3, 0)$ F) $\left(\sqrt{3}/2, -\sqrt{3}/2\right)$

   G) $(0, -3)$ H) $\left(-\sqrt{3}/2, \sqrt{3}/2\right)$

   Answer: $(-3, 0)$ easy

   _____

2. Convert the Cartesian coordinates $(1, 1)$ to polar coordinates.

   A) $(1, \pi/4)$ B) $\left(\sqrt{2}, \pi/4\right)$ C) $\left(\sqrt{2}, \pi/2\right)$

   D) $(1, 2\pi)$ E) $(1, \pi)$ F) $\left(\sqrt{2}, \pi\right)$

   G) $(1, \pi/2)$ H) $\left(\sqrt{2}, 2\pi\right)$

   Answer: $\left(\sqrt{2}, \pi/4\right)$ easy

   _____

3. Which of the three sets of polar coordinates below represent the point whose rectangular coordinates are $(0, 1)$?

   1) $(1, \pi/2)$

   2) $(-1, 3\pi/2)$

   3) $(1, 3\pi/2)$

   A) none B) 1 C) 2 D) 3

   E) 1, 2 F) 1, 3 G) 2, 3 H) 1, 2, 3

   Answer: 1, 2 easy

   _____

4. Find a polar equation for the curve represented by the Cartesian equation $x^2 + (y - 1)^2 = 1$.

   A) $r = \cos^2 \theta$ B) $r = 2 \sin^2 \theta$ C) $r = \sin^2 \theta$

   D) $r = 2 \sin \theta$ E) $r = 2 \cos^2 \theta$ F) $r = \cos \theta$

   G) $r = \sin \theta$ H) $r = 2 \cos \theta$

   Answer: $r = 2 \sin \theta$ medium

   _____

5.　Find a Cartesian equation for the curve represented by the polar equation

$r \sin \theta + r^2 \cos^2 \theta + r^2 = 0$.

A) $y + 2y^2 + 2x^2 = 0$ 　　B) $y^2 + x + x^2 = 0$ 　　C) $x^2 + y + y^2 = 0$

D) $x + 2x^2 + 2y^2 = 0$ 　　E) $x^2 + 2y + 2y^2 = 0$ 　　F) $2x + 2x^2 + y^2 = 0$

G) $2y^2 + x + x^2 = 0$ 　　H) $2x^2 + y + y^2 = 0$

Answer: $2x^2 + y + y^2 = 0$ medium

6.　Find the smallest positive value of θ for which the curve $r = 1 + \cos \theta$ has a horizontal tangent.

A) $7\pi/6$ 　　B) $2\pi/3$ 　　C) $\pi/2$ 　　D) $\pi/4$

E) $\pi/3$ 　　F) $3\pi/4$ 　　G) $\pi/6$ 　　H) $5\pi/6$

Answer: $\pi/3$ medium

7.　Express the polar equation $r = -4 \sin \theta$ in rectangular form.

Answer: $x^2 + (y + 2)^2 = 4$ medium

8.　Let P have rectangular coordinates $(1, 1)$. Find a set of polar coordinates (r, θ) for P such that $r > 0, \; \theta < 0$.

Answer: $\left(\sqrt{2}, \frac{-7\pi}{4}\right)$ medium

9.　Let P have rectangular coordinates $(1, 1)$. Find a set of polar coordinates (r, θ) for P such that $r < 0, \; \theta > 0$.

Answer: $\left(-\sqrt{2}, \frac{5\pi}{4}\right)$ medium

10.　Let P have rectangular coordinates $(1, 1)$. Find a set of polar coordinates (r, θ) for P such that $r < 0, \; \theta < 0$.

Answer: $\left(-\sqrt{2}, \frac{-3\pi}{4}\right)$ medium

11. Find polar coordinates for the Cartesian point $\left(-2, 2\sqrt{3}\,\right)$.

Answer: $\left(4, \frac{2\pi}{3}\right)$ easy

12. Find Cartesian coordinates for the point whose polar coordinates are $\left(-3, \frac{3\pi}{4}\right)$.

Answer: $\left(\frac{3}{\sqrt{2}}, \frac{-3}{\sqrt{2}}\right)$ easy

13. Find polar coordinates for the Cartesian point $(0, -5)$.

Answer: $\left(5, \frac{3\pi}{2}\right)$ easy

14. Convert $x^3 + xy^2 - y^2 = 0$ to polar form and write r explicitly in terms of θ.

Answer: $r = \dfrac{\sin^2 \theta}{\cos \theta}$ or $r = \sin \theta \tan \theta$ medium

Calculus, 2nd Edition
by James Stewart
Chapter 9, Section 5
Areas and Lengths in Polar Coordinates

1. Find the area of the region bounded by the curve $r = 4$ and lying in the sector $0 \leq \theta \leq \pi/2$.

    A) 8π B) 4π C) 2π D) $\pi/2$

    E) π F) 16π G) 12π H) 3π

    Answer: 4π easy

2. Find the area of the region bounded by the curve $r = 4\theta$ and lying in the sector $0 \leq \theta \leq \pi/2$.

    A) $\pi^2/12$ B) $\pi^2/24$ C) $\pi^2/6$ D) $\pi^3/3$

    E) $\pi^2/3$ F) $\pi^3/6$ G) $\pi^3/24$ H) $\pi^3/12$

    Answer: $\pi^3/3$ medium

3. Find the area of the region bounded by the curve $r = \sin\theta$.

    A) 4π B) $3\pi/2$ C) 2π D) $\pi/2$

    E) $\pi/4$ F) $\pi/6$ G) π H) $\pi/3$

    Answer: $\pi/4$ medium

4. Find the area of the region inside the curve $r = 4\sin\theta$ but not inside the curve $r = 2\sin\theta$.

    A) 12π B) π C) 4π D) $\pi/2$

    E) 2π F) 6π G) 8π H) 3π

    Answer: 3π medium

5. Find the area of the region that lies inside both the curves $r = 4\sin\theta$ and $r = 4\cos\theta$.

    A) $2\pi - 2$ B) $2\pi + 4$ C) $\pi - 2$ D) $\pi - 1$

    E) $\pi + 1$ F) $\pi + 2$ G) $2\pi + 2$ H) $2\pi - 4$

    Answer: $2\pi - 4$ hard

6. Find the length of the curve $r = 4 \sin \theta$.

A) 2π B) 4π C) 6π D) $\pi/2$

E) π F) 8π G) 12π H) 16π

Answer: 4π easy

7. Find the length of the curve $r = e^\theta$, $0 \le \theta \le 2\pi$.

A) $e^\pi + 1$ B) $e^{2\pi} + 1$ C) $e^{2\pi} - 1$

D) $\sqrt{2}\left(e^\pi - 1\right)$ E) $\sqrt{2}\left(e^{2\pi} - 1\right)$ F) $\sqrt{2}\left(e^{2\pi} + 1\right)$

G) $e^\pi - 1$ H) $\sqrt{2}\left(e^\pi + 1\right)$

Answer: $\sqrt{2}\left(e^{2\pi} - 1\right)$ medium

8. Find the area enclosed by one loop of the curve $r^2 = 2 \sin \theta$.

A) $1/2$ B) 1 C) $3/2$ D) 2

E) $5/2$ F) 3 G) $7/2$ H) 4

Answer: $1/2$ medium

9. Write the definite integral representing the length of the polar curve $r = f(\theta)$, $a \le \theta \le b$.

A) $\int_a^b r \frac{dr}{d\theta}\, d\theta$ B) $\int_a^b \left(r + \frac{dr}{d\theta}\right) d\theta$ C) $\int_a^b r^2 \frac{d^2 r}{d\theta^2}\, d\theta$

D) $\int_a^b \left(r^2 + \frac{d^2 r}{d\theta^2}\right) d\theta$ E) $\int_a^b \sqrt{1 + \frac{d^2 r}{d\theta^2}}\, d\theta$ F) $\int_a^b \sqrt{r^2 + \left(\frac{dr}{d\theta}\right)^2}\, d\theta$

G) $\int_a^b \sqrt{1 + r^2}\, d\theta$ H) $\int_a^b \sqrt{1 + \theta^2}\, d\theta$

Answer: $\int_a^b \sqrt{r^2 + \left(\frac{dr}{d\theta}\right)^2}\, d\theta$ easy

10. Find the area of the region that lies inside the circle $r = 3 \cos \theta$ and outside the cardioid $r = 1 + \cos \theta$.

A) $\pi/3$ B) $\pi/2$ C) $2\pi/3$ D) π

E) $4\pi/3$ F) $3\pi/2$ G) $5\pi/3$ H) 2π

Answer: π hard

11. Find the area of the region inside the polar curve given by $r = 2 - 2\sin\theta$ and outside the polar curve given by $r = 3$.

Answer: $\dfrac{9\sqrt{3}}{2} - \pi$ medium

12. Sketch the graph of $r = 2(\sin\theta + \cos\theta)$. Find the area of the region enclosed by the curve.

Answer:

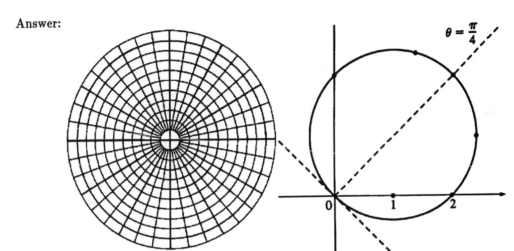

$\theta = \dfrac{\pi}{4}$

Area $= 2\pi$

13. Sketch the graphs of the circle $r = 6\cos\theta$ and the cardioid $r = 2 + 2\cos\theta$ on the same coordinate system. Find the area of the region that is inside the circle and outside the cardioid.

Answer:

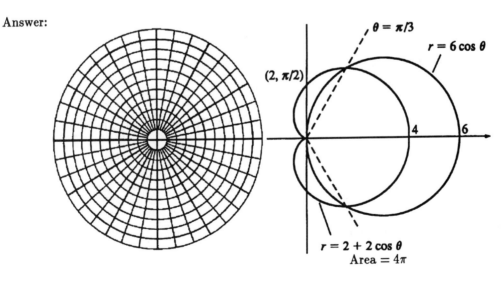

$\theta = \pi/3$

$r = 6\cos\theta$

$(2, \pi/2)$

$r = 2 + 2\cos\theta$
Area $= 4\pi$

14. Find the total area outside $r = 1$ and inside $r = 2 \sin 3\theta$.

Answer: $\dfrac{\sqrt{3}}{2} - \dfrac{\pi}{3}$ medium

15. Find the area outside $r = 2$ but inside $r = 4 \cos \theta$.

Answer: $\dfrac{4\pi}{3} + 2\sqrt{3}$ medium

16. Sketched below is the propeller (with equation $r = 2 \sin 3\theta$) to be mounted atop the beanies which will soon be required headgear for all mathematics, physics, and chemistry majors. As the proud wearer walks, the propeller will spin causing the words "mathematics," physics," and "chemistry" printed on the blades to blur and blend signifying the unity of pure science. Find the area of ONE BLADE of the propeller.

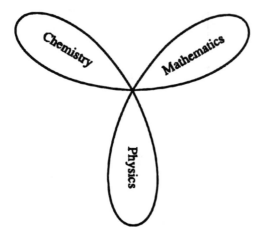

Answer: Area $= \dfrac{\pi}{3}$ medium

17. Find the area of the region R inside the circle $r = \sin \theta$ and outside the cardioid $r = 1 + \cos \theta$.

Answer: $\dfrac{4 - \pi}{4}$ hard

18. Calculate the total area enclosed by the leaves of $r = 5 \cos 3\theta$.

Answer: $\dfrac{25\pi}{4}$ medium

19. Sketch the bifolium $r = a \sin \theta \cos^2\theta$. Find the total area enclosed.

Answer:

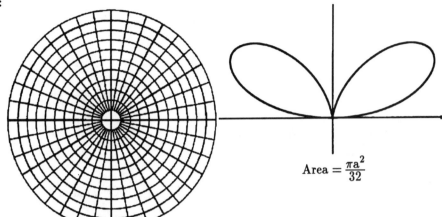

$$\text{Area} = \frac{\pi a^2}{32}$$

20. Sketch the graph and compute the area enclosed by the graph of the polar equation

$$r = 2 + \cos 2\theta$$

Answer:

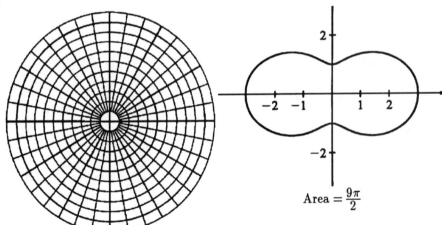

$$\text{Area} = \frac{9\pi}{2}$$

21. Find the area of the region inside the circle $r = 3$ and outside the spiral $r = \theta$, between the values of θ where $\theta = 0$ and where the two graphs intersect.

Answer: Area $= 9$ medium
_____

22. Find the area inside the curve $r = 4 \cos 2\theta$ and outside the curve $r = 2$.

Answer: $4\sqrt{3} + \frac{8\pi}{3}$ medium

23. Find the points of intersection of $r = 3 \cos \theta$ and $r = 3 - 3 \cos \theta$.

Answer: $\left(\frac{3}{2}, \frac{\pi}{3}\right)$, $\left(\frac{3}{2}, \frac{5\pi}{3}\right)$, $(0,0)$ medium

24. Find all points of intersection of the curve $r = 2 - 4 \sin \theta$ with the curve $r = 2 \sin \theta$.

Answer: $(0, 0)$, $\left(0, \frac{\pi}{6}\right)$, $\left(\frac{2}{3}, \sin^{-1}(1/3)\right)$, $\left(\frac{2}{3}, \pi - \sin^{-1}(1/3)\right)$ hard

25. Find the length of the curve $r = \sin^3\left(\frac{\theta}{3}\right)$.

Answer: $\pi + \frac{3\sqrt{3}}{8}$ medium

26. Find the arc length of the curve $r = 3 \sin \theta$.

Answer: 3π medium

Calculus, 2nd Edition
by James Stewart
Chapter 9, Section 6
Conic Sections

1.   Give the y-coordinate of the focus of the parabola $y = x^2$.

    A) 2   B) 4   C) $-1/4$ D) $-1/2$

    E) 1/2   F) -4 G) 1/4   H) -2

Answer: 1/4 easy

2.   Find the distance between focus and directrix of the parabola $y = x^2$.

    A) 2   B) 8   C) 16   D) 1/8

    E) 1/2   F) 1   G) 4   H) 1/4

Answer: 1/2 medium

3.   Find the distance between the two foci of the ellipse $\dfrac{x^2}{4} + \dfrac{y^2}{8} = 1$.

    A) 2   B) $2\sqrt{2}$ C) 8   D) $\sqrt{2}$

    E) 4   F) $4\sqrt{2}$ G) 1   H) $8\sqrt{2}$

Answer: 4 medium

4.   Find the distance between the two foci of the hyperbola $x^2 - y^2 = 8$.

    A) 2   B) 8   C) $2\sqrt{2}$ D) 4

    E) 1   F) $4\sqrt{2}$ G) $\sqrt{2}$ H) $8\sqrt{2}$

Answer: 8 medium

5.   Find the focus of the parabola $2y^2 - 3y = 4x + 5$.

    A) $(-33/32,\, 3/4)$ B) $(-33/32,\, 3/8)$ C) $(49/8,\, -5/4)$

    D) $(49/4,\, -5/8)$ E) $(-33/32,\, -3/4)$ F) $(-33/32,\, -3/8)$

    G) $(49/8,\, 5/4)$ H) $(49/4,\, 5/8)$

Answer: $(-33/32,\, 3/4)$ hard

6. An ellipse has foci located at $(-1, 0)$ and $(1, 0)$ and vertices located at $(-2, 0)$ and $(2, 0)$. How long is the vertical axis?

A) 2   B) 6   C) $2\sqrt{2}$ D) $4\sqrt{2}$

E) 3   F) $2\sqrt{3}$ G) 4   H) $4\sqrt{3}$

Answer: $2\sqrt{3}$ medium

7. A hyperbola has asymptotes $y = \pm\ x/2$ and vertices $(-1, 0)$ and $(1, 0)$. Find the distance between its foci.

A) $2\sqrt{3}$ B) $\sqrt{3}$ C) $\sqrt{5}$ D) $\sqrt{5}/4$

E) $\sqrt{5}/2$ F) $\sqrt{3}/4$ G) $\sqrt{3}/2$ H) $2\sqrt{5}$

Answer: $\sqrt{5}$ medium

8. Find an equation for the ellipse which has foci at the points $(2, 2)$ and $(2, -4)$, and is tangent to the line $y = 4$.

Answer: $\dfrac{(x-2)^2}{16} + \dfrac{(y+1)^2}{25} = 1$ medium

9. Sketch and discuss $4x^2 - 8x + 3y^2 + 6y + 7 = 0$.

Answer:

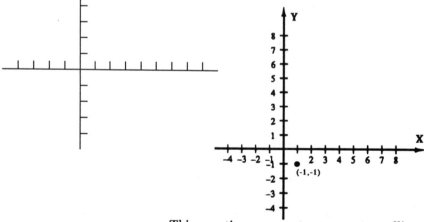

This equation appears to represent an ellipse because $A = 4$ and $B = 3$ are not equal and both are positive. Completing the square yields $4(x-1)^2 + 3(y+1)^2 = 0$. The only point that satisfies this equation is $(1, -1)$. This is known as a degenerate case of an ellipse.   medium

10. Use the definition of the ellipse to find the equation of the locus of a point the sum of whose distances from $(-3, -4)$ and $(3, 4)$ is 12.

Answer: $27x^2 - 24xy + 2y^2 = 396$ medium

11. Consider the ellipse given by $25x^2 + 4y^2 + 150x - 40y + 225 = 0$. Find the center, the vertices, and the foci of the ellipse.

Answer: center: $(-3, 5)$; vertices: $(-3, 0)$ and $(-3, 10)$; foci: $\left(-3, 5 + \sqrt{21}\right)$ and $\left(-3, 5 - \sqrt{21}\right)$

medium

12. Find an equation of the conic which has foci at the points $(3, 4)$ and $(3, 8)$, and has vertices at the points $(3, 3)$ and $(3, 9)$.

Answer: $\dfrac{(x-3)^2}{5} + \dfrac{(y-6)^2}{9} = 1$ medium

13. Use the definition of a parabola to find the equation of the parabola with focus $(-1, 3)$ and directrix $x = 5$.

Answer: $(y - 3)^2 = -12(x - 2)$ easy

14. Find the area bounded by the parabola $x^2 = 4py$ and the line $y = p$.

Answer: $\dfrac{8p^2}{3}$ easy

15. Sketch the following conic: $x^2 - 4x - y^2 - 6y = 4$.

Answer:

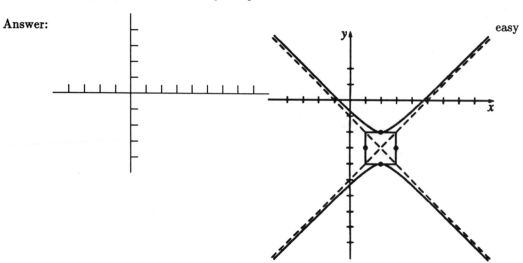

easy

16. Given the hyperbola $4x^2 + 16x - 3y^2 + 18y + 1 = 0$, name the center, vertices and equations of the asymptotes.

Answer: center: $(-2, 3)$; vertices: $(-2, 5)$ and $(-2, 1)$; asymptotes: $y = 3 \pm \dfrac{\sqrt{3}}{2}(x + 2)$ medium

17. Identify the following as a circle, parabola, ellipse, or hyperbola: $5x^2 + y^2 - 3x - 15 = 0$.

Answer: ellipse easy

18. Identify the following as a circle, parabola, ellipse, or hyperbola: $3x^2 - 2y + 5x + 1 = 0$.

Answer: parabola easy

19. Identify the following as a circle, parabola, ellipse, or hyperbola: $x + 3y + 2x^2 - 2y^2 = 0$.

Answer: hyperbola easy

20. Identify the following as a circle, parabola, ellipse, or hyperbola: $3 - 2x^2 + 2x - 2y^2 + 4y = 0$.

Answer: circle easy

21. Sketch the graph of $4x^2 - 8x + y^2 - 4y = 8$.

Answer:

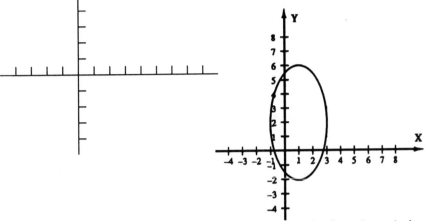

ellipse with center at $(1,2)$, major axis (parallel to y-axis) 8 units, minor axis 4 units medium

22. The vertical face of a dam is the lower half of an ellipse whose major axis is 60 feet long and forms the top of the dam. The minor axis is 40 feet long. Find the force exerted on the face of the dam if the water level is at the top of the dam.

Answer: $8000p$ lbs. where p lb/ft^3 is the weight density of the liquid. hard

Calculus, 2nd Edition
by James Stewart
Chapter 9, Section 7
Conic Sections in Polar Coordinates

1. An ellipse has axes of lengths 1 and 3. Find its eccentricity.

   A) 3   B) $\sqrt{8}/3$ C) $\sqrt{3}$ D) 2/3

   E) $\sqrt{3}/4$ F) $\sqrt{3}/2$ G) $\sqrt{2}$ H) 2

   Answer: $\sqrt{8}/3$ medium

2. An ellipse has eccentricity equal to $1/2$. Find the ratio of its major axis length $2a$ to its minor axis length $2b$.

   A) $3/\sqrt{2}$ B) $\sqrt{6}$ C) $\sqrt{2}$ D) $2/\sqrt{3}$

   E) $3\sqrt{2}$ F) $\sqrt{3}$ G) $2\sqrt{3}$ H) $2\sqrt{2}$

   Answer: $2/\sqrt{3}$ hard

3. Find the eccentricity of the ellipse whose polar equation is $r = \dfrac{1}{2 - \cos\theta}$.

   A) 1/4   B) 2   C) 1/2   D) 1/10

   E) 1   F) 1/8   G) 1/6   H) 1/12

   Answer: 1/2   easy

4. Find the major axis length $2a$ of the ellipse whose polar equation is $r = \dfrac{1}{2 - \cos\theta}$.

   A) 5/2   B) 2   C) 3/4   D) 3

   E) 8/3   F) 4/3   G) 2/3   H) 3/2

   Answer: 4/3   medium

5. Find the distance from vertex to focus for the parabola whose polar equation is $r = \dfrac{4}{1 - \cos\theta}$.

   A) 3   B) 2   C) 8   D) 1

   E) 4   F) 1/4   G) 1/3   H) 1/2

   Answer: 2   medium

6. Find the distance between the vertices of the hyperbola whose polar equation is $r = \dfrac{4}{1 - 2\cos\theta}$.

A) 8/3   B) 2   C) 6   D) 4

E) 4/3   F) 3   G) 2/3   H) 16/3

Answer: 8/3   medium

7. Find the eccentricity of the hyperbola whose polar equation is $r = \dfrac{4}{1 - 2\cos\theta}$.

A) 4   B) 8/3   C) 2   D) 4/3

E) 3   F) 6   G) 2/3   H) 16/3

Answer: 2   easy

8. Write the equation of the ellipse with eccentricity $\frac{2}{3}$ and foci $(-5, 4)$ and $(7, 4)$.

Answer: $\dfrac{(x-1)^2}{81} + \dfrac{(y-4)^2}{45} = 1$

9. Sketch the graph of $r = 3\cos\left(\theta - \frac{\pi}{4}\right)$.

Answer:

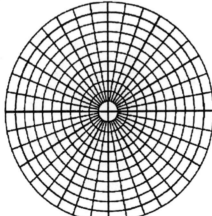 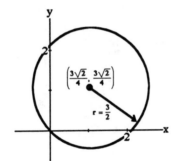

circle with center $\left(\dfrac{3\sqrt{2}}{4}, \dfrac{3\sqrt{2}}{4}\right)$ and radius $\frac{3}{2}$ medium

10.   Sketch the graph of $r = a \cos\theta + a \sin\theta$.

Answer:

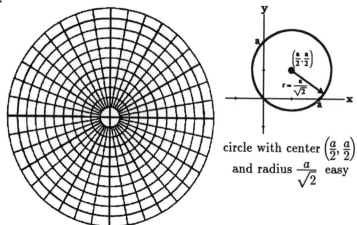

circle with center $\left(\dfrac{a}{2}, \dfrac{a}{2}\right)$
and radius $\dfrac{a}{\sqrt{2}}$ easy

11.   Identify and sketch the conic $r = \dfrac{10}{3 - 2\sin\theta}$.

Answer:

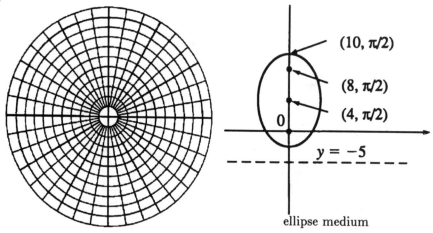

$(10, \pi/2)$
$(8, \pi/2)$
$(4, \pi/2)$
$y = -5$

ellipse medium

12.   Identify and sketch the conic $r = \dfrac{6}{1 + 5\cos\theta}$.

Answer:

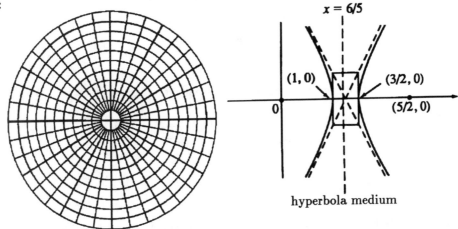

x = 6/5

(1, 0)   (3/2, 0)

0   (5/2, 0)

hyperbola medium

13.   Identify and sketch the conic $r = \dfrac{1}{1 - \sin\theta}$.

Answer:

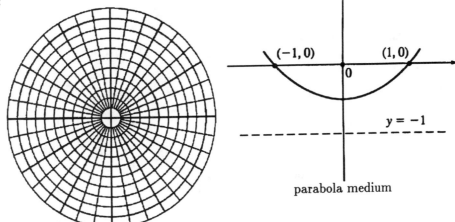

(−1, 0)   (1, 0)

0

y = −1

parabola medium

Calculus, 2nd Edition
by James Stewart
Chapter 10, Section 1
Sequences

1   Find a formula for the general term a_n of the sequence $\left\{1, -\frac{1}{2}, \frac{1}{4}, -\frac{1}{8}, \cdots\right\}$.

     A) 2^{1-n} B) 2^{n-1} C) $(-2)^n$ D) $(-2)^{n-1}$

     E) $(-1)^{2n}$ F) $(-2)^{1-n}$ G) $(-2)^{2n}$ H) 2^{-n}

     Answer: $(-2)^{1-n}$ easy

2.   Find a formula for the general term a_n of the sequence $\{1, 6, 120, 5040, \ldots\}$.

     A) $3^n(n+1)!$ B) $3^n \, n!$ C) $(n+1)!$ D) $(n+2)!$

     E) $(2n)!$ F) $n!$ G) $(2n-1)!$ H) $2^n \, n!$

     Answer: $(2n-1)!$ medium

3.   Determine the limit of the sequence $a_n = (n!)^{-1}$.

     A) divergent   B) $\ln 2$ C) e^2 D) e

     E) 1 F) $\ln 3$ G) $e-1$ H) 0

     Answer: 0 easy

4.   Determine the limit of the sequence $a_n = \dfrac{\sqrt{n+1}-\sqrt{n}}{\sqrt{n+1}+\sqrt{n}}$.

     A) $1/4$ B) $\sqrt{2}$ C) 4 D) $1/\sqrt{2}$

     E) 0 F) divergent   G) 2 H) $1/2$

     Answer: 0 medium

5.   Determine the limit of the sequence $a_n = (n+1)!/n!$.

     A) 2 B) 1 C) $1/3$ D) $1/2$

     E) divergent   F) 3 G) e H) 0

     Answer: divergent (easy)

6. Find the limit of the sequence $a_n = e^n/n!$.

    A) $(e^2 - 1)/e$ B) $\sqrt{e}$ C) e D) e^2

    E) 0   F) $(e-1)/e$ G) divergent   H) 1

    Answer: 0 medium
    _____

7. Find the limit of the sequence $\left\{ \sqrt{3},\ \sqrt{3\sqrt{3}},\ \sqrt{3\sqrt{3\sqrt{3}}},\ ... \right\}$.

    A) 1   B) e^3 C) $e^{3/2}$ D) 3

    E) 1/3   F) $e^{\sqrt{3}}$ G) π H) divergent

    Answer: 3 hard
    _____

8. If $a_1 = 1$ and $a_{n+1} = \sqrt{1 + a_n}$ for $n \geq 1$, and $\lim_{n\to\infty} a_n = L$ is assumed to exist, then what must L be?

    A) $\sqrt{2}$ B) $\sqrt{3}$ C) $\sqrt{5}$ D) $\sqrt{7}$

    E) $(1 + \sqrt{2})/2$ F) $(2 + \sqrt{3})/4$ G) $(1 + \sqrt{5})/2$ H) $(3 + \sqrt{7})/2$

    Answer: $(1 + \sqrt{5})/2$ hard
    _____

9. Determine the limit of the sequence $a_n = (-1)^n/\sqrt{n}$.

    A) -1 B) 0   C) 1/2   D) 1

    E) $\sqrt{2}$ F) 2   G) e H) divergent

    Answer: 0 medium
    _____

10. Determine the limit of the sequence $a_n = (-2)^n/n$.

    A) -2 B) 0   C) ln 2   D) $\sqrt{2}$

    E) e^2 F) $-1/2$ G) 1   H) divergent

    Answer: divergent medium
    _____

11. Determine the limit of the sequence $a_n = \dfrac{5\cos n}{n}$.

    A) 0   B) 1   C) 2   D) 3

    E) 4   F) 5   G) 6   H) divergent

    Answer: 0 medium
    _____

12. Determine the limit of the sequence $a_n = 2^n/n!$.

A) 0   B) 1   C) 2   D) 3

E) 4   F) 5   G) 6   H) divergent

Answer: 0 medium

_____

13. State the precise condition on r for the sequence $a_n = r^n$ to have a finite limit.

A) $r < 1$ B) $-1 < r < 1$ C) $-1 \leq r < 1$ D) $-1 < r \leq 1$

E) $-1 \leq r \leq 1$ F) $-1 < r$ G) $r \leq 1$ H) $-1 \leq r$

Answer: $-1 < r \leq 1$ medium

_____

14. Determine the limit of the sequence $a_n = \left[\ln(n+1) - \ln(n)\right]$.

A) $1/e$ B) 1   C) 2   D) 0

E) e F) $1/4$ G) $\ln 2$ H) divergent

Answer: 0 medium

_____

15. Determine the limit of the sequence $a_n = \dfrac{n!}{(n+3)!}$.

A) 0   B) 1   C) 2   D) e

E) 3   F) $1/2$ G) $1/3$ H) divergent

Answer: 0 medium

_____

16. Determine the limit of the sequence $a_n = \dfrac{\sin n}{\sqrt{n}}$.

A) 0   B) 1   C) 2   D) 3

E) 4   F) 5   G) $\sqrt{e}$ H) divergent

Answer: 0 medium

_____

17. If $a_1 = 1$ and $a_{n+1} = 3 - (1/a_n)$ for $n \geq 1$, find the limit of the sequence a_n.

A) 2   B) $\sqrt{2}$ C) $\sqrt{3}$ D) $\sqrt{5}$

E) $(2+\sqrt{3})/4$ F) $(3+\sqrt{5})/2$ G) $(5+\sqrt{7})/2$ H) $(5+2\sqrt{2})/3$

Answer: $(3+\sqrt{5})/2$ hard

_____

18. If $\frac{3n-1}{n+1} < x_n < \frac{3n^2+6n+2}{n^2+2n+1}$ for all positive integers n, then

A) $\lim\limits_{n\to\infty} x_n = L$ where $-1 \le L \le 2$ 　　　B) $\lim\limits_{n\to\infty} x_n = 3$

C) $\{x_n\}$ must be divergent 　　　D) $\{x_n\}$ is monotonic

E) $\{x_n\}$ is bounded but may be divergent

Answer: $\lim\limits_{n\to\infty} x_n = 3$　medium

---

19. Show, using the definition of the limit of a sequence, that $\lim\limits_{n\to\infty} \frac{3}{n} = 0$.

Answer: We wish to show that for every $\epsilon > 0$ there is a positive integer N such that is $n > N$,

$\left|\frac{3}{n} - 0\right| < \epsilon$. Suppose $\epsilon > 0$. $\frac{3}{\epsilon} > 0$. By the theorem of Archimedes there is a positive

integer N such that $N > \frac{3}{\epsilon}$. Now if $n > N$, $n > N > \frac{3}{\epsilon}$; whence, by elementary algebra

we have $\frac{1}{n} < \frac{\epsilon}{3}$. So if $n > N$ then $\left|\frac{3}{n} - 0\right| = \frac{3}{n} = 3\left(\frac{1}{n}\right) < 3\left(\frac{\epsilon}{3}\right) = \epsilon$. Hence by the

definition of limit, $\lim\limits_{n\to\infty} \frac{3}{n} = 0$. 　hard

---

20. The sequence defined by $x_n = \frac{(n+1)(n+3)}{n^2}$ is

A) decreasing 　　　B) increasing 　　　C) non-monotonic

D) divergent 　　　E) bounded above but not bounded below

Answer: decreasing 　medium

---

21. Consider the recursive sequence defined by $x_1 = 1$; $x_{n+1} = \frac{x_n^2 + 2}{2x_n}$, $n > 1$.

Evaluate the first three terms of this sequence.

Answer: $x_1 = 1$; $x_2 = \frac{3}{2}$; $x_3 = \frac{17}{12}$　easy

---

22. Consider the recursive sequence defined by $x_1 = 1$; $x_{n+1} = \dfrac{x_n^2 + 2}{2x_n}$, $n > 1$.

You may assume the sequence to be monotonic (after the first term) and bounded and hence convergent. Find its limit.

Answer: $L = \sqrt{2}$ medium
_____

23. Write the first five terms of the sequence: $a_n = \dfrac{4n - 3}{3n + 4}$.

Answer: $\left\{ \dfrac{1}{7}, \dfrac{1}{2}, \dfrac{9}{13}, \dfrac{13}{16}, \dfrac{17}{19}, \ldots \right\}$ easy
_____

24. Write the first five terms of the sequence: $a_n = \left\{ \dfrac{(-7)^{n+1}}{n!} \right\}$.

Answer: $\left\{ 49, -\dfrac{343}{2}, \dfrac{2401}{6}, -\dfrac{16807}{24}, \dfrac{117649}{120}, \ldots \right\}$ easy
_____

25. Find a formula for the general term a_n of $\left\{ \dfrac{1}{2}, \dfrac{1}{42}, \dfrac{1}{6}, \dfrac{1}{8}, \ldots \right\}$ assuming the pattern of the first few terms continues.

Answer: $a_n = \dfrac{1}{2n}$ easy
_____

26. Find a formula for the general term a_n of $\left\{ \dfrac{3}{2}, -\dfrac{9}{4}, \dfrac{27}{8}, -\dfrac{81}{16}, \ldots \right\}$ assuming the pattern of the first few terms continues.

Answer: $a_n = (-1)^{n+1} \left(\dfrac{3}{2} \right)^n$ medium
_____

27. Determine whether $a_n = 4\sqrt{n}$ converges or diverges. If it converges, find the limit.

Answer: diverges easy
_____

28. Determine whether $a_n = \sin(n\pi/2)$ converges or diverges. If it converges, find the limit.

Answer: diverges medium

29. Determine whether $\left\{ \dfrac{n!}{(n+2)!} \right\}$ converges or diverges. If it converges, find the limit.

Answer: converges to 0 medium

30. Determine whether $a_n = \dfrac{n \cos n}{n^2 + 1}$ converges or diverges. If it converges, find the limit.

Answer: converges to 0 medium

31. Determine whether $a_n = \dfrac{1}{5^n}$ is increasing, decreasing, or not monotonic.

Answer: decreasing easy

32. Determine whether $a_n = \dfrac{3n+4}{2n+5}$ is increasing, decreasing, or not monotonic.

Answer: increasing medium

33. Determine whether $a_n = 3 + (-1)^n/n$ is increasing, decreasing, or not monotonic.

Answer: not monotonic medium

34. Determine whether $a_n = \dfrac{\sqrt{n+1}}{5n+3}$ is increasing, decreasing, or not monotonic.

Answer: decreasing hard

Calculus, 2nd Edition
by James Stewart
Chapter 10, Section 2
Series

1.　Find the sum of the series $1 + \frac{1}{3} + \frac{1}{9} + \frac{1}{27} + \cdots$.

A) 7/4　　　　B) 11/6　　　　C) 3/2　　　　D) 4/3

E) 5/3　　　　F) 7/6　　　　G) 5/4　　　　H) divergent

Answer: 3/2　easy

_____

2.　Find the sum of the series $2 + \frac{1}{2} + \frac{1}{8} + \frac{1}{32} + \cdots$.

A) 15/7　　　　B) 8/3　　　　C) 7/3　　　　D) 13/6

E) 16/7　　　　F) 5/2　　　　G) 17/6　　　　H) divergent

Answer: 8/3　medium

_____

3.　Find the sum of the series $\displaystyle\sum_{n=1}^{\infty} \frac{1}{n(n+2)}$.

A) 3/4　　　　B) 1/2　　　　C) 3/5　　　　D) 9/10

E) 7/10　　　　F) 4/5　　　　G) 2/3　　　　H) divergent

Answer: 3/4　hard

_____

4.　Express the number 1.363636 ... as a ratio of integers.

A) 17/13　　　　B) 31/19　　　　C) 30/19　　　　D) 15/13

E) 17/11　　　　F) 22/17　　　　G) 15/11　　　　H) 21/17

Answer: 15/11　medium

_____

5.　Find the values of x for which the series $\displaystyle\sum_{n=1}^{\infty} (x-1)^n$ converges.

A) $0 < x \le 2$　　　　B) $-2 \le x < 0$　　　　C) $-2 < x \le 0$

D) $0 \le x < 2$　　　　E) $-2 < x < 0$　　　　F) $0 < x < 2$

G) $0 \le x \le 2$　　　　H) $-2 \le x \le 0$

Answer: $0 < x < 2$ medium

_____

6.   A rubber ball is dropped from a height of 10 feet and bounces to 3/4 its height after each fall. If it continues to bounce until it comes to rest, find the total distance in feet it travels.

A) 55   B) 70   C) 40   D) 65
E) 35   F) 45   G) 50   H) 60

Answer: 70 medium
_____

7.   Find the sum of the series $\displaystyle\sum_{n=1}^{\infty} \frac{(-1)^{n-1}}{2^n}$.

A) 1/3   B) 1/2   C) 3/5   D) 1
E) 3/2   F) 2   G) 3   H) divergent

Answer: 1/3 medium
_____

8.   Find the sum of the series $\displaystyle\sum_{n=1}^{\infty} (-1)^{n+1} \frac{2^n}{3^n}$.

A) 1/3   B) 1/2   C) 3/5   D) 1
E) 3/2   F) 2/5   G) 3   H) divergent

Answer: 2/5 medium
_____

9.   Find the sum of the series $\displaystyle\sum_{n=0}^{\infty} \frac{1}{2^n}$.

A) −1   B) 0   C) 1/2   D) 1
E) 2   F) 3/2   G) 4/3   H) 3

Answer: 2 easy
_____

10.   Find the sum of the series $\displaystyle\sum_{n=1}^{\infty} \frac{(-2)^{n+3}}{5^{n-2}}$.

A) 400   B) 400/3   C) 80   D) 400/7
E) 400/9   F) 400/11   G) 400/13   H) divergent

Answer: 400/7 hard
_____

11.   Find the sum of the series $\displaystyle\sum_{n=1}^{\infty} \frac{(-3)^{n-1}}{4^n}$.

A) −3/4   B) −1/4   C) −1/7   D) 1/7
E) 1/4   F) 3/4   G) 0   H) divergent

Answer: 1/7 medium
_____

12. Find the sum of the series $\displaystyle\sum_{n=1}^{\infty} \frac{2^n}{3^{n-1}}$.

    A) 4/3   B) 5/3   C) 2   D) 3

    E) 4   F) 5   G) 6   H) divergent

    Answer: 6 medium

13. Find the sum of the series $\displaystyle\sum_{n=4}^{\infty} \frac{3}{n(n-1)}$.

    A) 1/4   B) 1/3   C) 1/2   D) 2/3

    E) 3/4   F) 1   G) 2   H) divergent

    Answer: 1/3 medium

14. Find the sum of the following series: $\displaystyle\sum_{n=1}^{\infty} \frac{2^n}{5^{n+1}}$

    Answer: $\frac{2}{15}$ medium

15. $\frac{2}{9} - \frac{4}{27} + \cdots + \frac{(-1)^{n+1} \cdot 2^n}{3^{n+1}} + \cdots$ is equal to

    A) 2/15   B) 1/5   C) 1/6   D) 21/110

    E) 1/7

    Answer: 2/15 medium

16. Find the value of $\displaystyle\sum_{n=2}^{\infty} \frac{3^n + 5^n}{15^n}$

    Answer: $S = \frac{13}{60}$ medium

17. Determine whether the series $1 - \frac{1}{2} + \frac{1}{4} - \frac{1}{8} + \cdots$ is convergent or divergent. If it is convergent, find its sum.

    Answer: converges with sum $\frac{2}{3}$ medium

18. Determine whether the series $\frac{1}{2^6} + \frac{1}{2^8} + \frac{1}{2^{10}} + \frac{1}{2^{12}} + \cdots$ is convergent or divergent. If it is convergent, find its sum.

    Answer: converges with sum $\frac{1}{48}$ medium

19. Determine whether the series $-\frac{81}{100} + \frac{9}{10} - 1 + \frac{10}{9} - \cdots$ is convergent or divergent. If it is convergent, find its sum.

Answer: diverges   medium

20. Determine whether the series $\sum\limits_{n=1}^{\infty} \frac{1}{e^{2n}}$ is convergent or divergent. If it is convergent, find its sum.

Answer: converges to $\frac{1}{e^2 - 1}$ medium

21. Determine whether the series $\sum\limits_{n=0}^{\infty} \frac{4^{n+1}}{5^n}$ is convergent or divergent. If it is convergent, find its sum.

Answer: converges to 20   medium

22. Determine whether the series $\sum\limits_{n=1}^{\infty} (-1)^{n-1} \frac{3^{2n}}{2^{3n+1}}$ is convergent or divergent. If it is convergent, find its sum.

Answer: diverges   medium

23. Determine whether the series $\sum\limits_{n=1}^{\infty} \frac{n^2}{3(n+1)(n+2)}$ is convergent or divergent. If it is convergent, find its sum.

Answer: diverges by the Test for Divergence   medium

Calculus, 2nd Edition
by James Stewart
Chapter 10, Section 3
The Integral Test

1. Find the sum of the series $\displaystyle\sum_{n=1}^{\infty} \frac{1}{\sqrt{n}}$.

   A) 20/3   B) e^2 C) 3   D) 20/7

   E) 2   F) 10/7   G) 10/3   H) divergent

   Answer: divergent easy
   _____

2. If we use the integral test to show that the series $\displaystyle\sum_{n=1}^{\infty} \frac{1}{n^2}$ converges, we obtain an improper integral with a finite value. What is that value?

   A) 2/3   B) 7/6   C) 3/2   D) 1/2

   E) 5/6   F) 4/3   G) 3/4   H) 1

   Answer: 1 easy
   _____

3. If we use the integral test to show that the series $\displaystyle\sum_{n=1}^{\infty} \frac{1}{n(n+1)}$ converges, we obtain an improper integral with a finite value. What is that value?

   A) e B) $\ln 2$ C) $\sqrt{3}$ D) $\sqrt{2}$

   E) $e-1$ F) $1/e$ G) $\ln 3$ H) 1

   Answer: ln 2 hard
   _____

4. Which of the three series below converge?

   1) $\displaystyle\sum_{n=1}^{\infty} \frac{1}{n}$ 2) $\displaystyle\sum_{n=1}^{\infty} \frac{1}{n^{1.1}}$ 3) $\displaystyle\sum_{n=1}^{\infty} \frac{1}{n^{0.9}}$

   A) 1, 2   B) 1   C) none   D) 3

   E) 2   F) 2, 3   G) 1, 2, 3   H) 1, 3

   Answer: 2 easy
   _____

5. Which of the three series below converge?

1) $\sum_{i=2}^{\infty} \dfrac{1}{n \ln n}$ 2) $\sum_{i=2}^{\infty} \dfrac{1}{n(\ln n)^2}$ 3) $\sum_{i=2}^{\infty} \dfrac{1}{n(\ln n)^3}$

A) 2 B) 1, 2, 3 C) 1, 3 D) 2, 3

E) 3 F) none G) 1, 2 H) 1

Answer: 2, 3 medium

6. According to the proof of the integral test, the sum of the series $\sum_{n=1}^{\infty} \dfrac{1}{n^{1.001}}$ must lie between what two values?

A) 101, 102 B) 102, 103 C) 999, 1000

D) 1001, 1002 E) 1002, 1003 F) 1000, 1001

G) 99, 100 H) 100, 101

Answer: 1000, 1001 hard

7. What is the value of p that marks the boundary between convergence and divergence of the series $\sum_{n=2}^{\infty} \dfrac{1}{n(\ln n)^p}$?

A) 1/2 B) 1/3 C) diverges for all p

D) ln 2 E) 1 F) ln 3

G) converges for all p H) 1/e

Answer: 1 medium

8. Which of three series below converge?

1) $\sum_{n=1}^{\infty} \dfrac{1}{n^4}$ 2) $\sum_{n=1}^{\infty} \dfrac{n^2}{2n^2+1}$ 3) $\sum_{n=1}^{\infty} (\sqrt{2})^n$

A) none B) 1 C) 2 D) 3

E) 1, 2 F) 1, 3 G) 2, 3 H) 1, 2, 3

Answer: 1 easy

9. The series $\displaystyle\sum_{n=1}^{\infty} \frac{1}{n^{\alpha}}$ converges if and only if

A) $\alpha < 1$ B) $-1 < \alpha < 1$ C) $\alpha \leq 1$

D) $\alpha > 1$ E) $\alpha \geq 1$ F) $\alpha < -1$

G) $\alpha < -1$ H) $-1 < \alpha < 1$

Answer: $\alpha > 1$ medium

10. Which of the three series below converge?

1) $\displaystyle\sum_{n=1}^{\infty} \frac{n}{n+1}$ 2) $\displaystyle\sum_{n=1}^{\infty} \frac{\pi^{n}}{3^{n}}$ 3) $\displaystyle\sum_{n=1}^{\infty} \frac{1}{n\sqrt{n}}$

A) none   B) 1   C) 2   D) 3

E) 1, 2   F) 1, 3   G) 2, 3   H) 1, 2, 3

Answer: 3   easy

11. Use the integral test to determine if the following series converges or diverges:

$$\sum_{n=1}^{\infty} \frac{n}{\left(n^2+1\right)^2}$$

Answer: the integral has a value of $\frac{1}{4}$, which, since it is finite, the series converges   easy

12. Use the integral test to show that the series $\displaystyle\sum_{k=2}^{\infty} \frac{1}{k(\ln k)^{p}}$ converges is $p > 1$ and diverges if

$p \leq 1$.   [Hint: Consider two cases $p = 1$ and $p \neq 1$.]

Answer: Suppose that $p = 1$. Using the integral test with $\displaystyle\int_{2}^{\infty} \frac{dx}{x \ln k}$ gives

$\displaystyle\lim_{l \to +\infty} \Big(\ln(\ln x)\Big)_{2}^{1} = \infty$ and the series diverges. If $p \neq 1$, then using the integral test

with $\displaystyle\int_{2}^{\infty} \frac{dx}{x(\ln x)^{p}}$ gives $\displaystyle\lim_{l \to \infty} \left(\frac{(\ln x)^{1-p}}{1-p}\right)_{2}^{1}$. Since $\ln x \to \infty$ as $x \to \infty$, the

convergence of the series depends on whether $\ln x$ is in the numerator or denominator of

the limit above. If $p > 1$, the $\ln x$ is in the denominator and the series converges. If

$p > 1$, the $\ln x$ is in the numerator and the series diverges. So we have convergence if

$p > 1$ and divergence if $p \leq 1$.   hard

13. Use the integral test to determine whether $\displaystyle\sum_{n=2}^{\infty} \frac{1}{n(\log n)^4}$ converges or diverges.

Answer: converges   medium
_____

14. Determine whether or not the following infinite series converges: $\displaystyle\sum_{k=2}^{\infty} \frac{1}{k(\ln k)^2}$

Answer: converges   medium
_____

15. Determine whether the following series converges or diverges. $\displaystyle\sum_{n=1}^{\infty} 3ne^{-n^2}$

Answer: converges   medium
_____

Calculus, 2nd Edition
by James Stewart
Chapter 10, Section 4
The Comparison Tests

1. Which of the following three tests will establish that the series $\sum_{n=1}^{\infty} \dfrac{3}{n(n+2)}$ converges?

    1) Comparison test with $\sum_{n=1}^{\infty} 3n^{-2}$

    2) Limit comparison test with $\sum_{n=1}^{\infty} n^{-2}$

    3) Comparison test with $\sum_{n=1}^{\infty} 3n^{-1}$

A) none   B) 1   C) 2   D) 3

E) 1, 2   F) 1, 3   G) 2, 3   H) 1, 2, 3

Answer: 1,2 medium

2. Which of the following three tests will establish that the series $\sum_{n=1}^{\infty} \dfrac{n}{\sqrt{2n^5+1}}$ converges?

    1) Comparison test with $\sum_{n=1}^{\infty} n^{-5/2}$

    2) Comparison test with $\sum_{n=1}^{\infty} n^{-3/2}$

    3) Comparison test with $\sum_{n=1}^{\infty} n^{-1/2}$

A) none   B) 1   C) 2   D) 3

E) 1, 2   F) 1, 3   G) 2, 3   H) 1, 2, 3

Answer: 2 easy

3. Which of the following three tests will establish that the series $\sum_{n=1}^{\infty} \frac{n}{\sqrt{7n^3 + 46}}$ diverges?

    1) Limit comparison test with $\sum_{n=1}^{\infty} n^{-1}$

    2) Comparison test with $\sum_{n=1}^{\infty} n^{-1}$

    3) Comparison test with $\sum_{n=1}^{\infty} n^{-1/2}$

A) none   B) 1   C) 2   D) 3

E) 1, 2   F) 1, 3   G) 2, 3   H) 1, 2, 3

Answer: 1   hard

4. Which of the following series converge?

    1) $\sum_{n=1}^{\infty} \frac{1}{n^2 + \sqrt{n}}$ 2) $\sum_{n=1}^{\infty} \frac{\sqrt{n}}{n^2 + \ln n}$ 3) $\sum_{n=1}^{\infty} \frac{n}{\sqrt{n^3 + 2n^2}}$

A) none   B) 1   C) 2   D) 3

E) 1, 2   F) 1, 3   G) 2, 3   H) 1, 2, 3

Answer: 1, 2   medium

5. Which of the following series converge?

    1) $\sum_{n=1}^{\infty} n^{-n}$ 2) $\sum_{n=1}^{\infty} e^{100-n}$ 3) $\sum_{n=1}^{\infty} \frac{n^n}{n!}$

A) none   B) 1   C) 2   D) 3

E) 1, 2   F) 1, 3   G) 2, 3   H) 1, 2, 3

Answer: 1, 2   medium

6. Which of the following series converge?

    1) $\sum_{n=1}^{\infty} \frac{1}{\ln(n+1)}$ 2) $\sum_{n=1}^{\infty} \frac{1}{[\ln(n+1)]^2}$ 3) $\sum_{n=1}^{\infty} \frac{1}{[\ln(n+1)]^3}$

A) none   B) 1   C) 2   D) 3

E) 1, 2   F) 1, 3   G) 2, 3   H) 1, 2, 3

Answer: none   medium

7.   Which of the following series converge?

   1) $\sum_{n=1}^{\infty} \dfrac{1}{e^n}$ 2) $\sum_{n=1}^{\infty} \dfrac{1}{\sqrt{e^n}}$ 3) $\sum_{n=1}^{\infty} \dfrac{1}{\sqrt[3]{e^n}}$

   A) none   B) 1   C) 2   D) 3
   E) 1, 2   F) 1, 3   G) 2, 3   H) 1, 2, 3

   Answer: 1, 2, 3 easy
   _____

8.   Which of the following series converge?

   1) $\sum_{n=1}^{\infty} \dfrac{3^{2n}}{2^{3n}}$ 2) $\sum_{n=1}^{\infty} \dfrac{1}{(n+1)^3}$ 3) $\sum_{n=1}^{\infty} \dfrac{n+1}{\sqrt{n^3+2}}$

   A) none   B) 1   C) 2   D) 3
   E) 1, 2   F) 1, 3   G) 2, 3   H) 1, 2, 3

   Answer: 2 medium
   _____

9.   Which of the following series converge?

   1) $\sum_{n=1}^{\infty} (-1)^n$ 2) $\sum_{n=1}^{\infty} 2^n$ 3) $\sum_{n=1}^{\infty} \dfrac{1}{2+n^3}$

   A) none   B) 1   C) 2   D) 3
   E) 1, 2   F) 1, 3   G) 2, 3   H) 1, 2, 3

   Answer: 3 easy
   _____

10.   Which of the following series converge?

   1) $\sum_{n=1}^{\infty} \dfrac{n}{\sqrt{n^3+n^2}}$ 2) $\sum_{n=2}^{\infty} \dfrac{n}{n\sqrt{n}-1}$ 3) $\sum_{n=1}^{\infty} \dfrac{1}{n^2+n+1}$

   A) none   B) 1   C) 2   D) 3
   E) 1, 2   F) 1, 3   G) 2, 3   H) 1, 2, 3

   Answer: 3 medium
   _____

11. Determine whether $\displaystyle\sum_{n=1}^{\infty} \frac{\cos n + 3^n}{n^2 + 5^n}$ is convergent or divergent.

Answer: convergent   medium

12. Determine whether the series $\displaystyle\sum_{n=0}^{\infty} \frac{1 + \sin^2 n}{5^n}$ converges.

Answer: converges   medium

13. Consider the two series: a) $\displaystyle\sum_{k=2}^{\infty} \frac{\ln k}{k}$ and b) $\displaystyle\sum_{k=2}^{\infty} \frac{1}{k \ln k}$. Suppose you compare (a) and (b) to the series $\displaystyle\sum_{k=1}^{\infty} \frac{1}{k}$. What (if anything) can you conclude about the convergence or divergence of (a) and (b) using ONLY this comparison test?

Answer: $\displaystyle\sum_{k=1}^{\infty} \frac{1}{k}$ diverges to ∞. Since $\ln k > 1$ for $k \geq 3$, we have a) $\dfrac{\ln k}{k} > \dfrac{1}{k}$ and

b) $\dfrac{1}{k \ln k} < \dfrac{1}{k}$. From a) we conclude that $\displaystyle\sum_{k=2}^{\infty} \frac{\ln k}{k}$ also diverges to ∞. b) nothing

can be concluded from b) above. The comparison test yields no useful information

about the series $\displaystyle\sum_{k=2}^{\infty} \frac{1}{k \ln k}$. hard

14. For the series below, tell whether or not it converges, and indicate what test you used. If the test involves a limit, give the limit. If the test involves a comparison, give the comparison.
$$\sum_{n=2}^{\infty} \frac{n^{1/n}}{\ln(n)}$$

Answer: It diverges by the comparison test, since $\displaystyle\sum_{n=2}^{\infty} \frac{1}{n}$ diverges, so does the given series. hard

Calculus, 2nd Edition
by James Stewart
Chapter 10, Section 5
Alternating Series

1. Which of the following are alternating series?

   1) $\displaystyle\sum_{n=1}^{\infty} \cos n$ 2) $\displaystyle\sum_{n=1}^{\infty} \cos n\pi$ 3) $\displaystyle\sum_{n=1}^{\infty} \sin n\pi$

   A) none   B) 1   C) 2   D) 3

   E) 1, 2   F) 1, 3   G) 2, 3   H) 1, 2, 3

   Answer: 2 easy

2. Which of the following series converge?

   1) $\displaystyle\sum_{n=1}^{\infty} \frac{(-1)^n}{\ln(n+1)}$ 2) $\displaystyle\sum_{n=1}^{\infty} (-1)^n \ln(n+1)$ 3) $1 - \frac{1}{2} + \frac{2}{3} - \frac{3}{4} + \frac{4}{5} - \frac{5}{6} + \cdots$

   A) none   B) 1   C) 2   D) 3

   E) 1, 2   F) 1, 3   G) 2, 3   H) 1, 2, 3

   Answer: 1 medium

3. If we add the first 100 terms of the alternating series $1 - \frac{1}{2} + \frac{1}{3} - \frac{1}{4} + \frac{1}{5} - \cdots$, how close can we determine the partial sum s_{100} to be to the sum s of the series?

   A) $s_{100} > s$, with $s_{100} - s < 1/101$ B) $s_{100} > s$, with $s_{100} - s < 1/e^{100}$

   C) $s_{100} > s$, with $s_{100} - s < 1/100$ D) $s_{100} < s$, with $s - s_{100} < 1/e^{101}$

   E) $s_{100} > s$, with $s_{100} - s < 1/e^{101}$ F) $s_{100} < s$, with $s - s_{100} < 1/101$

   G) $s_{100} < s$, with $s - s_{100} < 1/100$ H) $s_{100} < s$, with $s - s_{100} < 1/e^{100}$

   Answer: $s_{100} < s$, with $s - s_{100} < 1/100$ (medium)

4. How many terms of the alternating series $\sum_{n=1}^{\infty} (-1)^{n+1} n^{-2}$ must we add in order to be sure that the partial sum s_n is within 0.0001 of the true sum s?

A) 10   B) 300   C) 30000   D) 30

E) 3   F) 10000   G) 100   H) 1000

Answer: 100   medium

5. Which of the following series converge?

1) $\sum_{n=1}^{\infty} \frac{n}{n+1}$ 2) $\sum_{n=1}^{\infty} \frac{\sqrt{n+1}}{n^2+2}$ 3) $\sum_{n=1}^{\infty} \frac{(-1)^{n-1}}{\sqrt{n+1}}$

A) none   B) 1   C) 2   D) 3

E) 1, 2   F) 1, 3   G) 2, 3   H) 1, 2, 3

Answer: 2, 3   medium

6. Which of the following series converge?

1) $\sum_{n=1}^{\infty} \frac{1}{n}$ 2) $\sum_{n=1}^{\infty} \frac{1}{n^2}$ 3) $\sum_{n=1}^{\infty} \frac{(-1)^n}{n}$

A) none   B) 1   C) 2   D) 3

E) 1, 2   F) 1, 3   G) 2, 3   H) 1, 2, 3

Answer: 2, 3   easy

7. Which one of the following series diverges?

A) $\sum_{n=1}^{\infty} \left(\frac{3}{\pi}\right)^n$ B) $\sum_{n=2}^{\infty} \frac{1}{\sqrt{n^3+1}}$ C) $\sum_{n=4}^{\infty} \frac{(-1)^n}{\ln n}$

D) $\sum_{n=2}^{\infty} \frac{3}{n \ln n}$ E) $\sum_{n=1}^{\infty} \left(\frac{1}{n} - \frac{1}{n+1}\right)$ F) $\sum_{n=1}^{\infty} \left(\frac{2}{e}\right)^n$

G) $\sum_{n=1}^{\infty} \frac{3}{n^2 \ln n}$ H) $\sum_{n=1}^{\infty} 3n^{-3/2}$

Answer: $\sum_{n=2}^{\infty} \frac{3}{n \ln n}$ hard

8. Test the following series for convergence or divergence: $-5 - \frac{5}{2} + \frac{5}{5} - \frac{5}{8} + \frac{5}{11} - \frac{5}{14} + \cdots$

Answer: converges by the Alternating Series Test   easy

9. Test the following series for convergence or divergence: $\frac{1}{\ln 2} - \frac{1}{\ln 3} + \frac{1}{\ln 4} - \frac{1}{\ln 5} + \frac{1}{\ln 6} - \cdots$

Answer: converges by the Alternating Series Test   easy

10. Test the following series for convergence or divergence: $\displaystyle\sum_{n=1}^{\infty} \frac{(-1)^n}{\sqrt{n+3}}$

Answer: converges by the Alternating Series Test   medium

11. Test the following series for convergence or divergence: $\displaystyle\sum_{n=2}^{\infty} \frac{(-1)^{n-1}}{n \ln n}$

Answer: converges by the Alternating Series Test   medium

12. Test the following series for convergence or divergence: $\displaystyle\sum_{n=1}^{\infty} (-1)^n \frac{n^2}{n^2+1}$

Answer: diverges by the Divergence Test   medium

13. Test the following series for convergence or divergence: $\displaystyle\sum_{n=1}^{\infty} (-1)^{n+1} \frac{n}{2^n}$

Answer: converges by the Alternating Series Test   medium

14. Test the following series for convergence or divergence: $\sum_{n=1}^{\infty} (-1)^{n-1} \frac{\ln n}{n}$

Answer: converges by the Alternating Series Test medium

15. Test the following series for convergence or divergence: $\sum_{n=1}^{\infty} \frac{\sin\left(\frac{n\pi}{2}\right)}{n!}$

Answer: converges medium

16. Test the following series for convergence or divergence: $\sum_{n=1}^{\infty} (-1)^{n-1} \frac{(n+9)(n+10)}{n(n+1)}$

Answer: diverges medium

17. Test the following series for convergence or divergence: $\sum_{n=1}^{\infty} (-1)^n \cos\left(\frac{\pi}{n}\right)$

Answer: diverges medium

18. Test the following series for convergence or divergence: $\sum_{n=1}^{\infty} \frac{(-1)^n}{|n-10\pi|}$

Answer: converges medium

19. Approximate the sum $\sum_{n=1}^{\infty} \frac{(-1)^{n+1}}{n^4}$ with error < 0.001.

Answer: 0.948 medium

20. Approximate the sum $\sum_{n=0}^{\infty} \frac{(-1)^n n}{4^n}$ with error < 0.002.

Answer: -0.161 medium

1. Examine the two series below for absolute convergence (A), conditional convergence (C), or divergence (D).

   1) $\sum_{n=1}^{\infty} (-1)^n$ 2) $\sum_{n=1}^{\infty} (-1)^{n-1} n^{-1}$

   A) 1A, 2A   B) 1A, 2C   C) 1A, 2D   D) 1C, 2A

   E) 1C, 2C   F) 1C, 2D   G) 1D, 2A   H) 1D, 2C

   Answer: 1D, 2C   easy

2. Examine the two series below for absolute convergence (A), conditional convergence (C), or divergence (D).

   1) $\sum_{n=1}^{\infty} (-1)^{n-1} n^{-1}$ 2) $\sum_{n=1}^{\infty} (-1)^{n-1} n^{-2}$

   A) 1A, 2A   B) 1A, 2C   C) 1A, 2D   D) 1C, 2A

   E) 1C, 2C   F) 1C, 2D   G) 1D, 2A   H) 1D, 2C

   Answer: 1C, 2A   easy

3. Which of the following series will, when rearranged, converge to different values?

   1) $\sum_{n=1}^{\infty} n^{-1}$ 2) $\sum_{n=1}^{\infty} (-1)^{n-1} n^{-1}$ 3) $\sum_{n=1}^{\infty} (-1)^{n-1} n^{-2}$

   A) none   B) 1   C) 2   D) 3

   E) 1, 2   F) 1, 3   G) 2, 3   H) 1, 2, 3

   Answer: 2   medium

4. Which of the following series are conditionally convergent?

1) $\displaystyle\sum_{n=1}^{\infty} (-e)^{-n}$ 　　2) $\displaystyle\sum_{n=1}^{\infty} (-1)^{-n} n^{-1}$ 　　3) $\displaystyle\sum_{n=1}^{\infty} (-1)^{-n} n^{-2}$

A) none 　　　　B) 1 　　　　C) 2 　　　　D) 3

E) 1, 2 　　　　F) 1, 3 　　　　G) 2, 3 　　　　H) 1, 2, 3

Answer: 2 easy

_____

5. Examine the two series below for absolute convergence (A), conditional convergence (C), or divergence (D).

1) $\displaystyle\sum_{n=1}^{\infty} (-1)^{n-1} \frac{(n+2)3^n}{2^{2n+1}}$ 　　2) $\displaystyle\sum_{n=1}^{\infty} (-1)^{n-1} \frac{(n+3)2^{2n}}{3^n+100}$

A) 1A, 2A 　　　B) 1A, 2C 　　　C) 1A, 2D 　　　D) 1C, 2A

E) 1C, 2C 　　　F) 1C, 2D 　　　G) 1D, 2A 　　　H) 1D, 2C

Answer: 1A, 2D medium

_____

6. Examine the two series below for absolute convergence (A), conditional convergence (C), or divergence (D).

1) $\displaystyle\sum_{n=1}^{\infty} \frac{(-1)^{n-1}}{\ln(n+1)}$ 　　2) $\displaystyle\sum_{n=1}^{\infty} \frac{(-1)^{n-1}}{(\ln(n+1))^2}$

A) 1A, 2A 　　　B) 1A, 2C 　　　C) 1A, 2D 　　　D) 1C, 2A

E) 1C, 2C 　　　F) 1C, 2D 　　　G) 1D, 2A 　　　H) 1D, 2C

Answer: 1C, 2C medium

_____

7. Examine the two series below for absolute convergence (A), conditional convergence (C), or divergence (D).

1) $\displaystyle\sum_{n=1}^{\infty} (-1)^{n-1} \frac{n+1}{\ln(n+1)}$ 　　2) $\displaystyle\sum_{n=1}^{\infty} (-1)^{n-1} \frac{\ln(n+1)}{n+1}$

A) 1A, 2A 　　　B) 1A, 2C 　　　C) 1A, 2D 　　　D) 1C, 2A

E) 1C, 2C 　　　F) 1C, 2D 　　　G) 1D, 2A 　　　H) 1D, 2C

Answer: 1D, 2C medium

_____

8. Which of the following series converge?

1) $\sum_{n=1}^{\infty} \left(\frac{n}{2+3n}\right)^n$ 2) $\sum_{n=2}^{\infty} \frac{n+1}{\sqrt{n^4-1}}$ 3) $\sum_{n=1}^{\infty} \frac{1}{1+n^2}$

A) none   B) 1   C) 2   D) 3

E) 1, 2   F) 1, 3   G) 2, 3   H) 1, 2, 3

Answer: 1, 3 medium

9. Which of the following series are *conditionally* convergent?

1) $\sum_{n=1}^{\infty} \frac{(-1)^n}{n^2}$ 2) $\sum_{n=1}^{\infty} \frac{(-1)^{n-1}}{\sqrt{n}}$ 3) $\sum_{n=1}^{\infty} \frac{\cos n}{2^n}$

A) none   B) 1   C) 2   D) 3

E) 1, 2   F) 1, 3   G) 2, 3   H) 1, 2, 3

Answer: 2 medium

10. Which of the following series are *conditionally* convergent?

1) $\sum_{n=1}^{\infty} (-1)^{n+1} \frac{n+2}{n^2+1}$ 2) $\sum_{n=1}^{\infty} \frac{(-1)^n}{n^4}$ 3) $\sum_{n=1}^{\infty} \frac{\sin n\pi}{\pi^n}$

A) none   B) 1   C) 2   D) 3

E) 1, 2   F) 1, 3   G) 2, 3   H) 1, 2, 3

Answer: 1 medium

11. Which of the following series diverge?

1) $\sum_{n=1}^{\infty} \frac{n+2}{n^2+1}$ 2) $\sum_{n=1}^{\infty} \frac{n!}{2^n}$ 3) $\sum_{n=1}^{\infty} \left(\frac{2n-1}{n+3}\right)^n$

A) none   B) 1   C) 2   D) 3

E) 1, 2   F) 1, 3   G) 2, 3   H) 1, 2, 3

Answer: 1, 2, 3 hard

12. Which of the following series are absolutely convergent?

1) $\sum_{n=1}^{\infty} \frac{(-1)^n}{n^2}$ 2) $\sum_{n=1}^{\infty} \frac{(-1)^n}{n}$ 3) $\sum_{n=1}^{\infty} \frac{1}{n^3}$

A) none   B) 1   C) 2   D) 3
E) 1, 2   F) 1, 3   G) 2, 3   H) 1, 2, 3

Answer: 1, 3 easy

13. Which of the following series can be shown to be convergent *using the Ratio Test*?

1) $\sum_{n=1}^{\infty} \frac{1}{n^2}$ 2) $\sum_{n=1}^{\infty} \frac{n}{3^n}$ 3) $\sum_{n=1}^{\infty} \frac{2^n}{\sqrt{n}}$

A) none   B) 1   C) 2   D) 3
E) 1, 2   F) 1, 3   G) 2, 3   H) 1, 2, 3

Answer: 2 medium

14. Which *one* of the following series is divergent?

A) $\sum_{n=1}^{\infty} \left(\frac{2}{3}\right)^n$ B) $\sum_{n=1}^{\infty} \frac{1}{n^2+1}$ C) $\sum_{n=1}^{\infty} \frac{1}{n5^n}$

D) $\sum_{n=2}^{\infty} \frac{n}{n^2-1}$ E) $\sum_{n=1}^{\infty} \frac{n^3}{n^5+2}$ F) $\sum_{n=1}^{\infty} \frac{2^n}{n!}$

G) $\sum_{n=1}^{\infty} \frac{\pi}{n^2}$ H) $\sum_{n=1}^{\infty} (-1)^{n-1} \pi/n$

Answer: $\sum_{n=2}^{\infty} \frac{n}{n^2-1}$ hard

15. Determine whether the given series is conditionally convergent, absolutely convergent or

divergent. $\sum_{n=1}^{\infty} \frac{(-1)^{n+1}}{\sqrt{n}}$

Answer: conditionally convergent easy

16. Determine whether the given series is conditionally convergent, absolutely convergent or divergent.

$$\sum_{k=2}^{\infty} \frac{(-1)^{k+1}}{\ln k}$$

Answer: converges conditionally   medium

17. Determine whether the given series is conditionally convergent, absolutely convergent or divergent.

$$\sum_{n=1}^{\infty} \frac{(-1)^{n+1}n}{n^2+1}$$

Answer: converges conditionally   medium

18. Does the series $\sum_{n=1}^{\infty} \frac{2^{3n}}{5^n}$ converge or diverge? Justify.

Answer: $\frac{a_{n+1}}{a_n} = \frac{2^{3(n+1)}}{5^{n+1}} \cdot \frac{5^n}{2^{3n}} = \frac{2^{3n} \cdot 8 \cdot 5^n}{5^n \cdot 5 \cdot 2^{3n}} = \frac{8}{5} > 1$. Therefore by the ratio test this series

diverges. Or use the fact that it is a geometric series. medium

19. Test for convergence: $\sum_{n=1}^{\infty} \frac{3^n}{n!}$.

Answer: the series converges   medium

20. Test for convergence: $\sum_{n=0}^{\infty} \frac{n!}{3^n}$.

Answer: the series diverges   medium

21. Establish the convergence or divergence of the series $\sum_{n=1}^{\infty} \frac{(n+3)! - n!}{2^n}$.

Answer: divergent   medium

22. Which of the following series is convergent but not absolutely convergent?

A) $\displaystyle\sum_{n=1}^{\infty} \frac{1}{n}$ B) $\displaystyle\sum_{n=1}^{\infty} \frac{\sin n}{n^2}$ C) $\displaystyle\sum_{n=1}^{\infty} \frac{(-1)^n}{\sqrt{n}}$ D) $\displaystyle\sum_{n=1}^{\infty} \frac{3^n}{2^n + \sqrt{n}}$ E) $\displaystyle\sum_{n=1}^{\infty}$

$\dfrac{1-2n}{n+1}$

Answer: (c) is convergent by the alternating series test but not absolutely convergent

since $\displaystyle\sum_{0}^{\infty} \left| \frac{(-1)^n}{\sqrt{n}} \right| = \sum_{n=1}^{\infty} \frac{1}{n^{1/2}}$ is a divergent p-series. medium

23. Determine if the following infinite series converges absolutely, converges conditionally, or

diverges: $\displaystyle\sum_{k=1}^{\infty} \frac{(-1)^k k}{k^2 + 1}$

Answer: Since this is an alternating series, and $\displaystyle\lim_{k\to\infty} \frac{k}{k^2 + 1} = 0$, then the series must

converge. $\displaystyle\sum_{k=1}^{\infty} \left| \frac{(-1)^k k}{k^2 + 1} \right| = \sum_{k=1}^{\infty} \frac{k}{k^2 + 1}$ diverges since it is asymptotically proportional to

a divergent series, the harmonic series: $\displaystyle\lim_{k\to\infty} \frac{\frac{k}{k^2 + 1}}{\frac{1}{k}} = \lim_{k\to\infty} \frac{k^2}{k^2 + 1} = 1$. Therefore, the

given series converges conditionally. medium

Calculus, 2nd Edition
by James Stewart
Chapter 10, Section 7
Strategy for Testing Series

1. Use the Ratio Test to examine the two series below, stating: absolute convergence (A), divergence
(D), or Ratio Test inconclusive (I).

   1) $\displaystyle\sum_{n=1}^{\infty} n^{-100}$ 2) $\displaystyle\sum_{n=1}^{\infty} 100^{-n}$

   A) 1A, 2A   B) 1A, 2D   C) 1A, 2I   D) 1D, 2A

   E) 1D, 2D   F) 1D, 2I   G) 1I, 2A   H) 1I, 2D

   Answer: 1I, 2A easy
   _____

2. Use the Ratio Test to examine the two series below, stating: absolute convergence (A), divergence
(D), or Ratio Test inconclusive (I).

   1) $\displaystyle\sum_{n=1}^{\infty} (-1)^{n-1} \frac{2^{2n+1}}{5^n}$ 2) $\displaystyle\sum_{n=1}^{\infty} (-1)^{n-1} \frac{5^n}{2^{2n+1}}$

   A) 1A, 2A   B) 1A, 2D   C) 1A, 2I   D) 1D, 2A

   E) 1D, 2D   F) 1D, 2I   G) 1I, 2A   H) 1I, 2D

   Answer: 1A, 2D medium
   _____

3. For which of the following series will the Test for Divergence establish divergence?

   1) $\displaystyle\sum_{n=1}^{\infty} (-1)^n$ 2) $\displaystyle\sum_{n=1}^{\infty} n^{-1}$ 3) $\displaystyle\sum_{n=1}^{\infty} \frac{n+1}{2n}$

   A) none   B) 1   C) 2   D) 3

   E) 1, 2   F) 1, 3   G) 2, 3   H) 1, 2, 3

   Answer: 1, 3 medium
   _____

4. For which of the following series will the Ratio Test fail to give a definite answer (i.e., be inconclusive)?

   1) $\sum_{n=1}^{\infty} (99/100)^n$ 2) $\sum_{n=1}^{\infty} (100/99)^n$ 3) $\sum_{n=1}^{\infty} n^{-100}$

   A) none   B) 1   C) 2   D) 3
   E) 1, 2   F) 1, 3   G) 2, 3   H) 1, 2, 3

   Answer: 3 medium
   _____

5. Tell which of the following three series can be compared with geometric series to establish convergence.

   1) $\sum_{n=1}^{\infty} \frac{1}{2+3^n}$ 2) $\sum_{n=1}^{\infty} \frac{n}{n^3+4}$ 3) $\sum_{n=1}^{\infty} (-1)^{n-1} \frac{n^2}{3^n}$

   A) none   B) 1   C) 2   D) 3
   E) 1, 2   F) 1, 3   G) 2, 3   H) 1, 2, 3

   Answer: 1, 3 hard
   _____

6. Tell which of the following three series cannot be found convergent by the ratio test but can be found convergent by comparison with a p-series.

   1) $\sum_{n=1}^{\infty} \frac{1}{2+3^n}$ 2) $\sum_{n=1}^{\infty} \frac{n}{n^3+4}$ 3) $\sum_{n=1}^{\infty} (-1)^{n-1} \frac{n^2}{3^n}$

   A) none   B) 1   C) 2   D) 3
   E) 1, 2   F) 1, 3   G) 2, 3   H) 1, 2, 3

   Answer: 2 hard
   _____

7. Which *one* of the following series diverges?

   A) $\sum_{n=1}^{\infty} \frac{1}{n(2n+1)}$ B) $\sum_{n=1}^{\infty} \frac{2n}{n+1}$ C) $\sum_{n=1}^{\infty} \frac{1}{(n+1)(n+3)}$

   D) $\sum_{n=1}^{\infty} \frac{1}{3^n}$ E) $\sum_{n=1}^{\infty} \frac{n-2}{n2^n}$ F) $\sum_{n=1}^{\infty} \frac{2n}{n!}$

   G) $\sum_{n=1}^{\infty} \frac{n^{100}}{n!}$ H) $\sum_{n=1}^{\infty} \frac{n^{100}}{2^n}$

   Answer: $\sum_{n=1}^{\infty} \frac{2n}{n+1}$ hard
   _____

8.   Which of the following series converge?

1) $\sum_{n=1}^{\infty} \dfrac{1}{1+n^3}$ 2) $\sum_{n=1}^{\infty} \dfrac{(-1)^n}{\sqrt[3]{n}}$ 3) $\sum_{n=1}^{\infty} \left(\dfrac{3n+1}{2n+1}\right)^n$

A) none   B) 1   C) 2   D) 3
E) 1, 2   F) 1, 3   G) 2, 3   H) 1, 2, 3

Answer: 1, 2 medium

9.   Which of the following series converge?

1) $\sum_{n=1}^{\infty} n^{-0.9}$ 2) $\sum_{n=1}^{\infty} \dfrac{3^n}{n+5^n}$ 3) $\sum_{n=1}^{\infty} \dfrac{n}{1+4n}$

A) none   B) 1   C) 2   D) 3
E) 1, 2   F) 1, 3   G) 2, 3   H) 1, 2, 3

Answer: 2 medium

10.   Analyze the following series for convergence vs. divergence: $\sum_{n=1}^{\infty} \cos\left(\dfrac{1}{n^2}\right)$.

Answer: diverges medium

11.   Analyze the following series for convergence vs. divergence: $\sum_{n=1}^{\infty} \dfrac{1}{e^n}$.

Answer: converges easy

12.   Analyze the following series for convergence vs. divergence: $\sum_{n=1}^{\infty} \dfrac{3^n}{n!}$.

Answer: converges medium

13. Analyze the following series for convergence vs. divergence: $\displaystyle\sum_{n=1}^{\infty} \frac{\sin n}{n^2}$.

Answer: converges (absolutely) medium

_____

14. Analyze the following series for convergence vs. divergence: $\displaystyle\sum_{n=1}^{\infty} \frac{n}{5n^2 + \sqrt{n}}$.

Answer: diverges medium

_____

15. Analyze the following series for convergence vs. divergence: $\displaystyle\sum_{n=1}^{\infty} \frac{(-1)^n}{n}$.

Answer: converges easy

_____

1. Find the radius of convergence of $\sum_{n=0}^{\infty} 3x^n$.

   A) 2　　　　　B) 1/2　　　　C) 6　　　　D) 1/6

   E) 3　　　　　F) 1　　　　　G) 1/3　　　H) 0

   Answer: 1　easy
   _____

2. Find the radius of convergence of $\sum_{n=0}^{\infty} (3x)^n$.

   A) 3　　　　　B) 0　　　　　C) 2　　　　D) 1/6

   E) 6　　　　　F) 1/3　　　G) 1　　　　H) 1/2

   Answer: 1/3　easy
   _____

3. Find the interval of convergence of $\sum_{n=0}^{\infty} \frac{x^n}{3n+1}$.

   A) $[-3,3]$　　　B) $(-1,1)$　　　C) $(-3,3)$　　　D) $[-1,1]$

   E) $(-1,1]$　　　F) $(-3,3]$　　　G) $[-3,3)$　　　H) $[-1,1)$

   Answer: $[-1,1)$　medium
   _____

4. Find the interval of convergence of $\sum_{n=0}^{\infty} \frac{(-3x)^n}{3n+1}$.

   A) $(-1/3,1/3)$　　B) $(-1/3,1/3]$　　C) $[-1/3,1/3)$　　D) $[-1/3,1/3]$

   E) $(-3,3)$　　　F) $(-3,3]$　　　G) $[-3,3)$　　　H) $[-3,3]$

   Answer: $(-1/3,1/3]$　medium
   _____

5. Find the interval of convergence of $\sum_{n=0}^{\infty} \frac{x^n}{2n^2}$.

   A) $[-1,1]$　　　B) $(-1,1]$　　　C) $(-1,1)$　　　D) $[-1,1)$

   E) $[-2,2]$　　　F) $(-2,2]$　　　G) $(-2,2)$　　　H) $[-2,2)$

   Answer: $[-1,1]$　medium
   _____

6. Find the radius of convergence of $\displaystyle\sum_{n=0}^{\infty} \frac{(2n)!}{n!} x^n$.

A) 2   B) 1/e   C) 1   D) 0

E) 2e   F) e   G) 1/2   H) 1/(2e)

Answer: 0   hard
_____

7. Find the radius of convergence of $\displaystyle\sum_{n=0}^{\infty} x^n/n!$.

A) 1/e   B) ∞ C) 2   D) 0

E) 1/(2e)   F) 1/2   G) e   H) 2e

Answer: ∞ easy
_____

8. Find the radius of convergence of $\displaystyle\sum_{n=0}^{\infty} \frac{4^n}{n!}(x-2)^n$.

A) 0   B) 1/4   C) 1/2   D) 1

E) 2   F) 4   G) 8   H) ∞

Answer: ∞ medium
_____

9. Find the radius of convergence of $\displaystyle\sum_{n=1}^{\infty} \frac{(x+3)^n}{\sqrt{n}\, 2^n}$.

A) 1/2   B) 1   C) 2   D) 3

E) 4   F) 5   G) 6   H) ∞

Answer: 2   medium
_____

10. Find the interval of convergence of $\displaystyle\sum_{n=1}^{\infty} \frac{(x+3)^n}{\sqrt{n}\, 2^n}$.

A) $(-1,2)$ B) $(-2,2]$ C) $[-3,3)$ D) $[-3,3]$

E) $(-5,-1)$ F) $[-5,-1)$ G) $(-1,3)$ H) $[-1,3]$

Answer: $[-5,-1)$ medium
_____

11. Find the radius of convergence of $\displaystyle\sum_{n=1}^{\infty} \frac{(x+2)^n}{n^2 3^n}$.

A) 0   B) 1   C) 2   D) 3

E) 4   F) 5   G) 6   H) ∞

Answer: 3   medium
_____

12. Find the interval of convergence of $\sum_{n=1}^{\infty} \frac{(x+2)^n}{n^2 3^n}$.

A) $[-5,1]$ B) $(-3,3)$ C) $[-2,3)$ D) $[-\infty,\infty)$

E) $(-5,1)$ F) $[-1,5)$ G) $(-3,2]$ H) $[-2/3\ ,\ 2/3]$

Answer: $[-5,1]$ hard

---

13. Find the radius of convergence of $\sum_{n=0}^{\infty} \frac{n}{4^n}(x+3)^n$.

A) 0   B) 1   C) 2   D) 3

E) 4   F) 5   G) 6   H) ∞

Answer: 4 easy

---

14. Find the interval of convergence of $\sum_{n=0}^{\infty} \frac{n}{4^n}(x+3)^n$.

A) $(-3,3)$ B) $[-3,3)$ C) $[-3,3]$ D) $(-3,3]$

E) $(-7,1)$ F) $[-7,1)$ G) $[-7,1]$ H) $(-7,1]$

Answer: $(-7,1)$ medium

---

15. Find the interval of convergence of $\sum_{n=1}^{\infty} \frac{x^n}{n^2}$.

A) $(-\infty,\infty)$ B) $(-1,1)$ C) $(-1,1]$ D) $[-1,1)$

E) $[-1,1]$ F) $(-4,4)$ G) $(-4,4]$ H) $[-4,4)$

Answer: $[-1,1]$ medium

---

16. Find the radius of convergence of $\sum_{n=1}^{\infty} (-1)^n \sqrt{n}\ 4^n (x-6)^n$.

A) 0   B) 2   C) 4   D) 6

E) 1/2   F) 1/4   G) 1/6   H) ∞

Answer: 1/4 medium

---

17. Find the interval of convergence of $\sum_{n=1}^{\infty} \frac{x^n}{n}$.

A) $(-1,1)$ B) $(-1,1]$ C) $[-1,1)$ D) $[-1,1]$

E) $(-e,e)$ F) $(-e,e]$ G) $[-e,e)$ H) $[-e,e]$

Answer: $[-1,1)$ medium

18. Find the radius of convergence of $\sum_{n=0}^{\infty} \frac{(-1)^n \, x^{2n+1}}{2n+1}$.

   A) 1   B) 1/2   C) 2   D) 0

   E) ∞ F) 1/4   G) 4   H) 8

Answer: 1 medium

_____

19. Find the radius of convergence of $\sum_{n=1}^{\infty} \frac{(-1)^n (x+2)^n}{\sqrt{n} \, 3^n}$.

   A) 0   B) 1   C) 2   D) 3

   E) 4   F) 5   G) 6   H) ∞

Answer: 3 medium

_____

20. Find the interval of convergence of $\sum_{n=1}^{\infty} \frac{(-1)^n (x+2)^n}{\sqrt{n} \, 3^n}$.

   A) $(-\infty, \infty)$ B) $(-5, 1]$ C) $[-5, 1)$ D) $[-5, 1]$

   E) $(-4, 0]$ F) $[-4, 0)$ G) $[-4, 0]$ H) $(-3, -1)$

Answer: $(-5, 1]$ medium

_____

21. Find the radius of convergence of $\sum_{n=1}^{\infty} \frac{(x-2)^n}{n3^n}$.

   A) 0   B) 1   C) 2   D) 3

   E) 4   F) 5   G) 6   H) ∞

Answer: 3 medium

_____

22. Find the interval of convergence of $\sum_{n=1}^{\infty} \frac{(x-2)^n}{n3^n}$.

   A) $(1, 3)$ B) $[1, 3)$ C) $(2, 4]$ D) $[2, 4]$

   E) $(-1, 5)$ F) $[-1, 5)$ G) $(-2, 6)$ H) $(-\infty, \infty)$

Answer: $[-1, 5)$ medium

_____

23. Find the radius of convergence of $\sum_{n=0}^{\infty} \frac{(-3)^n (x-1)^n}{\sqrt{n+1}}$.

   A) 0   B) 1/3   C) 1/2   D) 1

   E) 2   F) 3   G) 4   H) ∞

Answer: 1/3 medium

_____

24. Find the interval of convergence of $\sum\limits_{n=0}^{\infty} \dfrac{(-3)^n (x-1)^n}{\sqrt{n+1}}$.

A) $(2/3, 4/3]$ B) $[2/3, 4/3)$ C) $(1/2, 3/2)$ D) $[1/2, 3/2]$

E) $(0, 2]$ F) $[0, 2)$ G) $(-1, 3)$ H) $[-2, 4]$

Answer: $(2/3, 4/3]$ hard
_____

25. Find the interval of convergence of the power series $\sum\limits_{k=0}^{\infty} \dfrac{(-1)^k k^2}{5^k} (x-2)^k$.

Answer: $(-3, 7)$ medium
_____

26. Find the interval of convergence for $\sum\limits_{n=1}^{\infty} \dfrac{x^n}{4n^2}$.

Answer: $[-1, 1]$ easy
_____

27. Find the interval of convergence for $\sum\limits_{n=1}^{\infty} \dfrac{x^n}{n2^n}$.

Answer: $[-2, 2)$ easy
_____

28. Find the interval of convergence of the power series $\sum\limits_{n=1}^{\infty} \dfrac{(-1)^n (x-2)^n}{\sqrt[3]{n}}$.

Answer: $(1, 3]$ medium
_____

29. Consider the power series $\sum\limits_{k=1}^{\infty} \dfrac{(-1)^k}{k} x^k$.

(a) Find the radius of convergence.

(b) Determine what happens at the end points (absolute or conditional convergence, or divergence.)

Answer: (a) $r = 1$; (b) for $x = 1$, converges conditionally; for $x = -1$, diverges   medium
_____

30. Which of the following is impossible for the convergence set of the power series $\displaystyle\sum_{n=0}^{\infty} a_n (x-3)^n$?

A) $0 < x < 6$ B) $1 \le x < 5$ C) $2 \le x \le 5$ D) $2 \le x \le 4$

E) $-1 < x \le 7$

Answer: $2 \le x \le 5$ medium

31. Find the radius of convergence of the series $\displaystyle\sum_{n=1}^{\infty} \frac{3^n (x-2)^{2n+1}}{n!}$.

Answer: $r = \infty$; i.e., the series converges for all real x. medium

32. Find the interval of convergence for $\displaystyle\sum_{n=0}^{\infty} x^n$.

Answer: $(-1, 1)$ easy

33. Find the interval of convergence for $\displaystyle\sum_{k=0}^{\infty} \left(\frac{e^k}{k+1} \right) x^k$.

Answer: $\frac{-1}{e} \le x \le \frac{1}{e}$

34. Find the interval of convergence for $\displaystyle\sum_{k=1}^{\infty} \frac{x^k}{2^k k^2}$.

Answer: $[-2, 2]$ medium

35. Find the interval of convergence for $\displaystyle\sum_{k=1}^{\infty} \frac{(-1)^k (x-3)^k}{5^k (k+1)}$.

Answer: $(-2, 8]$ medium

Calculus, 2nd Edition
by James Stewart
Chapter 10, Section 9
Taylor and Maclaurin Series

1. Find the coefficient of x^2 in the Maclaurin series for $f(x) = \dfrac{1}{x+2}$.

    A) 1   B) 1/8   C) 1/2   D) −1/4

    E) −1/2   F) −1   G) 1/4   H) −1/8

    Answer: 1/8   medium

2. Find the coefficient of x^3 in the Maclaurin series for $f(x) = \sin 2x$.

    A) −2/3   B) −4/3   C) 4/3   D) −8/3

    E) 2/3   F) 8/3   G) −1/3   H) 1/3

    Answer: −4/3   medium

3. Find the radius of convergence of the Maclaurin series for $f(x) = \dfrac{1}{4+x^2}$.

    A) 1   B) 1/8   C) ∞   D) 1/4

    E) 1/2   F) 4   G) 8   H) 2

    Answer: 2   medium

4. Find the coefficient of x^2 in the Maclaurin series for e^{x-1}.

    A) e B) 1/e   C) 1/(2e)   D) 2

    E) 1/2   F) 0   G) 1   H) $e/2$

    Answer: 1/(2e)   hard

5. Find the coefficient of x^2 in the Maclaurin series for $f(x) = e^{-x^2}$.

    A) 1/4   B) −1   C) 1/2   D) −2

    E) 1   F) −1/4   G) −1/2   H) 2

    Answer: −1   easy

6. Find the coefficient of x^5 in the Maclaurin series for $f(x) = \displaystyle\int \cos(x^2)\,dx$. (Note: the series is unique except for the constant of integration.)

A) $-1/10$ B) $1/15$ C) $-1/5$ D) $2/5$

E) $-2/5$ F) $-1/15$ G) $1/5$ H) $1/10$

Answer: $-1/10$ medium

_____

7. Find a series representation of $\displaystyle\int \frac{e^x}{x}\,dx$.

A) $\displaystyle\sum_{n=0}^{\infty} \frac{x^n}{(n+1)!} + C$

B) $\displaystyle\sum_{n=0}^{\infty} \frac{x^{n+1}}{n!} + C$

C) $\displaystyle\sum_{n=0}^{\infty} \frac{x^{n+1}}{(n+1)!} + C$

D) $\displaystyle\sum_{n=1}^{\infty} \frac{(-1)^n}{n} x^n + C$

E) $\ln|x| + \displaystyle\sum_{n=1}^{\infty} \frac{x^n}{n \cdot n!} + C$

F) $\ln|x| + \displaystyle\sum_{n=1}^{\infty} \frac{x^n}{(n+1)!} + C$

G) $\ln|x| + \displaystyle\sum_{n=1}^{\infty} \frac{n+1}{n!} x^n + C$

H) $\ln|x| + \displaystyle\sum_{n=1}^{\infty} \frac{(-1)^n}{n!} x^n + C$

Answer: $\ln|x| + \displaystyle\sum_{n=1}^{\infty} \frac{x^n}{n \cdot n!} + C$ medium

_____

8. Find the terms in the Maclaurin series for the function $f(x) = \ln(1+x)$, as far as the term in x^3.

A) $1 - x + x^2 - x^3$

B) $x - x^2 + x^3$

C) $1 - x + \frac{1}{2}x^2 - \frac{1}{6}x^3$

D) $x - \frac{1}{2}x^2 + \frac{1}{3}x^3$

E) $1 + \frac{1}{2}x + \frac{2}{3}x^2 + \frac{5}{6}x^3$

F) $x + \frac{1}{2}x^2 + \frac{1}{6}x^3$

G) $1 + \frac{x}{2} + \frac{1}{6}x^2 + \frac{1}{24}x^3$

H) $x - \frac{1}{24}x^2 + \frac{1}{120}x^3$

Answer: $x - \frac{1}{2}x^2 + \frac{1}{3}x^3$ medium

_____

9. Find the terms in the Maclaurin series for the function $f(x) = e^{-x}$, as far as the term in x^3.

A) $1 - x + \frac{1}{2}x^2 - \frac{1}{6}x^3$

B) $1 + x + \frac{1}{2}x^2 - \frac{1}{6}x^3$

C) $1 - x + x^2 - x^3$

D) $1 + x + x^2 + x^3$

E) $1 - x + \frac{1}{2}x^2 - \frac{1}{3}x^3$

F) $1 + x + \frac{1}{2}x^2 + \frac{1}{3}x^3$

G) $-x + x^3$

H) $x - x^3$

Answer: $1 - x + \frac{1}{2}x^2 - \frac{1}{6}x^3$ easy

_____

10. Find the first four terms in the Maclaurin series for $f(x) = xe^{-x}$.

A) $x - x^2 + x^3 - x^4$ B) $x - \frac{1}{2}x^2 + \frac{1}{3}x^3 - \frac{1}{4}x^4$ C) $x - x^2 + \frac{1}{2}x^3 - \frac{1}{6}x^4$

D) $x - 2x^2 + 3x^3 - 4x^4$ E) $x + \frac{1}{2}x^2 + \frac{1}{6}x^3 + \frac{1}{24}x^4$ F) $x + x^2 + \frac{1}{3}x^3 + \frac{1}{8}x^4$

G) $x + x^2 + \frac{1}{2}x^3 + \frac{1}{6}x^4$ H) $\frac{1}{2}x - \frac{1}{6}x^2 + \frac{1}{24}x^3 - \frac{1}{120}x^4$

Answer: $x - x^2 + \frac{1}{2}x^3 - \frac{1}{6}x^4$ medium

_____

11. Use a Maclaurin series to approximate $\int_0^1 e^{-t^2} \, dt$ with an accuracy of 0.01.

Answer: 0.7428571   medium

_____

12. In terms of powers of x, the power series for $\frac{1}{1-x} = \sum_{n=0}^{\infty} x^n$. Find the power series for $\frac{1}{(1-x)^2}$ in terms of powers of x.

Answer: $\sum_{n=1}^{\infty} nx^{n-1}$ easy

_____

13. Find the Taylor series for $x \cos x$ about the origin.

Answer: $\sum_{n=0}^{\infty} \frac{(-1)^n x^{2n+1}}{(2n)!}$ hard

_____

14. Give the Taylor series expansion of $f(x) = \sin x$ about the point $c = \frac{\pi}{4}$.

Answer: $\frac{\sqrt{2}}{2} + \frac{\sqrt{2}}{2}\left(x - \frac{\pi}{4}\right) - \frac{\sqrt{2}}{2} \cdot \frac{1}{2!}\left(x - \frac{\pi}{4}\right)^2 - \frac{\sqrt{2}}{2} \cdot \frac{1}{3!}\left(x - \frac{\pi}{4}\right)^3 + \cdots$ medium

_____

15. Find the Taylor series for $y = \ln x$ at 2.

Answer: $\ln 2 + \sum_{n=1}^{\infty} \frac{(-1)^{n-1}}{n}\left(\frac{x-2}{2}\right)^n$ medium

_____

16. Find the Taylor polynomial of degree 4 at 0 for the function defined by $f(x) = \ln(1+x)$. Then compute the value of $\ln(1.1)$ accurate to as many decimal places as the polynomial of degree 4 allows.

Answer: $\ln(1+x) = x - \frac{1}{2}x^2 + \frac{1}{3}x^3 - \frac{1}{4}x^4$; $\ln(1.1) = .095$, accurate to three decimal places

medium

The following three questions pertain to the Taylor series about $x = 0$ for $f(x) = e^x$.

17. Derive the Taylor series about $x = 0$ for $f(x) = e^x$.

Answer: $\displaystyle\sum_{k=0}^{\infty} \frac{x^k}{k!}$ easy

18. Use the result of the last question to obtain the series expansion for e^{-x^2}.

Answer: $1 - x^2 + \frac{x^4}{2!} - \frac{x^6}{3!} + \frac{x^8}{4!} - \cdots$ medium

19. Use the result of the last question to obtain $\displaystyle\int_0^1 e^{-x^2}\, dx$ to two decimal place accuracy.

Answer: .74 medium

20. Find a Taylor series of degree 4 about $x = 0$ for $f(x) = \log \sec x$.

Answer: $\frac{x^2}{2} + \frac{x^4}{12}$ medium

21. Find the Maclaurin series expansion with $n = 5$ for $f(x) = 2^x$. Use this expansion to approximate $2^{.1}$.

Answer: $1 + (\ln 2)\, x + \frac{(\ln 2)^2 x^2}{2!} + \frac{(\ln 2)^3 x^3}{3!} + \frac{(\ln 2)^4 x^4}{4!} + \frac{(\ln 2)^5 x^5}{5!} + \cdots$; $2^{.1} \doteq 1.0718$ hard

22. Find the Maclaurin series expansion for $f(x) = \ln(1-x)$ and determine the interval of convergence.

    Answer: $\displaystyle\sum_{n=1}^{\infty} -\frac{1}{n}x^n$; converges for all x such that $-1 \leq x < 1$. medium

23. If the Maclaurin series for $f(x)$ is $1 - 9x + 16x^2 - 25x^3 + \cdots$, then $f^3(0)$ is equal to

    A) -25 B) $-25/6$ C) -150 D) $-25/3$

    E) -75

    Answer: -150 medium

24. Find the Taylor series of degree 5 about $x = 0$ for $y = \tan^2 x$.

    Answer: $x^2 + \frac{2}{3}x^4 + \cdots$ medium

Calculus, 2nd Edition
by James Stewart
Chapter 10, Section 10
The Binomial Series

1. Find the coefficient of x^3 in the binomial series for $(1+x)^5$.

   A) 3   B) 6   C) 15   D) 20

   E) 10   F) 5   G) 16   H) 12

   Answer: 10 easy
   _____

2. Find the coefficient of x in the binomial series for $\sqrt{1+x}$.

   A) 2   B) -1 C) 1   D) $-1/2$

   E) $1/2$ F) $-\sqrt{2}$ G) -2 H) $\sqrt{2}$

   Answer: $1/2$ easy
   _____

3. Find the coefficient of x^3 in the binomial series for $\sqrt{1+x}$.

   A) $-1/2$ B) $-1/4$ C) $1/8$ D) $1/2$

   E) $-1/8$ F) $-1/16$ G) $1/16$ H) $1/4$

   Answer: $1/16$ medium
   _____

4. Find the coefficient of x^3 in the binomial series for $\dfrac{1}{(1+x)^4}$.

   A) 6   B) 20   C) -6 D) -10

   E) -20 F) -12 G) 10   H) 12

   Answer: -20 medium
   _____

5. Use the binomial series to expand the function $\sqrt{4+x}$ as a power series. Give the coefficient of x^2 in that series.

   A) $-1/8$ B) $-1/32$ C) $-1/64$ D) $1/8$

   E) $1/32$ F) $-1/16$ G) $1/64$ H) $1/16$

   Answer: $-1/64$ medium
   _____

6. How many coefficients in the binomial series (expansion) of $(1 + x)^7$ are divisible by 7?

A) 0   B) 5   C) 7   D) 3

E) 2   F) 6   G) 1   H) 4

Answer: 6   medium

7. Find the terms in the power series expansion for the function $f(x) = \dfrac{1}{\sqrt{1+x^2}}$, as far as the term in x^3.

A) $1 - x + x^2 - x^3$ B) $x - \frac{1}{2}x^2 + \frac{1}{3}x^3$ C) $1 - x + \frac{1}{2}x^2 - \frac{1}{6}x^3$

D) $1 - x^2$ E) $1 + x^2$ F) $1 - \frac{1}{2}x + \frac{1}{24}x^2 - \frac{1}{120}x^3$

G) $1 - \frac{1}{2}x^2$ H) $x + \frac{1}{6}x^3$

Answer: $1 - \frac{1}{2}x^2$ medium

8. Find the terms of the Maclaurin (binomial) series for $\dfrac{1}{\sqrt{1-x}}$, as far as the term in x^3.

A) $1 - x + x^2 - x^3$ B) $1 - \frac{1}{2}x + \frac{1}{4}x^2 + \frac{1}{8}x^3$ C) $1 + \frac{1}{2}x + \frac{3}{8}x^2 + \frac{5}{16}x^3$

D) $1 - \frac{1}{2}x + \frac{3}{4}x^2 - \frac{5}{8}x^3$ E) $1 + \frac{1}{2}x + \frac{1}{4}x^2 + \frac{1}{6}x^3$ F) $1 - \frac{1}{2}x + \frac{1}{6}x^2 + \frac{1}{24}x^3$

G) $1 + \frac{1}{2}x + \frac{3}{4}x^2 + \frac{15}{16}x^3$ H) $1 - \frac{1}{2}x + \frac{3}{8}x^2 + \frac{7}{24}x^3$

Answer: $1 + \frac{1}{2}x + \frac{3}{8}x^2 + \frac{5}{16}x^3$ medium

9. Find the terms of the Maclaurin (binomial) series for $f(x) = \dfrac{1}{\sqrt{1+2x}}$, as far as the term in x^3.

A) $1 - x + x^2 - x^3$ B) $1 + x - \frac{1}{2}x^2 + \frac{1}{3}x^3$ C) $1 - x + \frac{3}{2}x^2 - \frac{5}{2}x^3$

D) $1 + x + 3x^2 + 5x^3$ E) $1 - x + \frac{3}{2}x^2 - \frac{7}{3}x^3$ F) $1 + x + \frac{1}{2}x^2 + \frac{7}{3}x^3$

G) $1 - x + \frac{5}{2}x^2 - \frac{7}{3}x^3$ H) $1 + x + \frac{7}{2}x^2 + \frac{11}{3}x^3$

Answer: $1 - x + \frac{3}{2}x^2 - \frac{5}{2}x^3$ hard

10. Express $\dfrac{1}{\sqrt{1+x}}$ as a power series in x and from the result obtain a binomial series for $\dfrac{1}{\sqrt{1-x^2}}$.

Answer: $1 - \frac{1}{2}x + \displaystyle\sum_{n=2}^{\infty} \frac{(-1)^n \cdot 1 \cdot 3 \cdot 5 \cdots (2n-1) x^n}{2^n \cdot n!}$; $1 + \frac{1}{2}x^2 + \displaystyle\sum_{0}^{\infty} \frac{1 \cdot 3 \cdot 5 \cdots (2n-1) x^{2n}}{2^n \cdot n!}$ hard

11. Use the binomial series formula to obtain the Maclaurin series for $f(x) = (1+x)^{1/3}$.

Answer: $1 + \frac{1}{3}x + \sum_{n=2}^{\infty} \frac{(-1)^{n+1} \cdot 2 \cdot 5 \cdot 8 \cdots (3n-4)\, x^n}{3^n \cdot n!}$, for $|x| < 1$ hard

12. Find a power series representation for $(b+x)^{3/2}$ where b is a perfect square, and state the radius of convergence in terms of b.

Answer: $b^{3/2} + \frac{3b^{1/2}x}{2} + 3\sum_{n=2}^{\infty} \frac{(-1)^n \cdot 1 \cdot 3 \cdot 5 \cdots (2n-5)\, x^n b^{(3-2n)/2}}{2^n \cdot n!}$; $r = b$ hard

13. Use the binomial series to expand $\frac{1}{(1+x)^3}$ as a power series. State the radius of convergence.

Answer: $1 + \sum_{n=1}^{\infty} \frac{(-1)^n 3 \cdot 4 \cdot 5 \cdots (n+2)\, x^n}{n!}$ with $R = 1$ medium

14. Use the binomial series to expand $\sqrt[3]{1+x^2}$ as a power series. State the radius of convergence.

Answer: $1 + \frac{x^2}{3} + \sum_{n=2}^{\infty} \frac{(-1)^{n-1} 2 \cdot 5 \cdot 8 \cdots (3n-4)\, x^{2n}}{3^n\, n!}$ with $R = 1$ medium

15. Use the binomial series to expand $\frac{1}{\sqrt{2+x}}$ as a power series. State the radius of convergence.

Answer: $\frac{\sqrt{2}}{2}\left[1 + \sum_{n=1}^{\infty} \frac{(-1)^n 1 \cdot 3 \cdot 5 \cdots (2n-1)\, x^n}{2^{2n}\, n!} \right]$ with $\left|\frac{x}{2}\right| < 1$ so $|x| < 2$ and $R = 2$.   hard

16. Use the binomial series to expand $(4+x)^{3/2}$ as a power series. State the radius of convergence.

Answer: $8 + 3x + \sum_{n=2}^{\infty} \frac{(3)(1)(-1) \cdots (5-2n)\, x^n}{8^{n-1}\, n!}$ with $\left|\frac{x}{4}\right| < 1$ so $|x| < 4$ and $R = 4$.   hard

17. Use the binomial series to expand $\frac{x^2}{\sqrt{1-x^3}}$ as a power series. State the radius of convergence.

Answer: $x^2 + \sum_{n=1}^{\infty} \frac{1 \cdot 3 \cdot 5 \cdots (2n-1)\, x^{3n+2}}{2^n\, n!}$ with $R = 1$ hard

18.   Use the binomial series to expand $\sqrt[5]{x-1}$ as a power series. State the radius of convergence.

Answer: $-1+\dfrac{x}{5}+\displaystyle\sum_{n=2}^{\infty}\dfrac{4\cdot 9\cdots(5n-6)\,x^n}{5^n\,n!}$ with $R=1$ hard

Calculus, 2nd Edition
by James Stewart
Chapter 10, Section 11
Approximation by Taylor Polynomials

1. According to Taylor's Formula, what is the maximum error possible in the use of the sum $\sum_{n=0}^{4} x^n/n!$ to approximate e^x in the interval $-1 \le x \le 1$?

   A) $e/240$ B) $e/48$ C) $e/480$ D) $e/24$

   E) $e/20$ F) $e/120$ G) $e/12$ H) $e/60$

   Answer: $e/120$ medium

   _____

2. Find the coefficient of $(x-2)^2$ in the Taylor polynomial $T_2(x)$ for the function x^3 at the number 2.

   A) 3   B) 0   C) 1   D) 6

   E) 2   F) 5   G) 8   H) 4

   Answer: 6 medium

   _____

3. What is the smallest value of n that will guarantee (according to Taylor's Formula) that the Taylor polynomial T_n at the number 0 will be within 0.0001 of e^x for $0 \le x \le 1$?

   A) 4   B) 5   C) 8   D) 6

   E) 7   F) 2   G) 3   H) 9

   Answer: 7 hard

   _____

4. Find the fourth degree Taylor's polynomial of the function $f(x) = xe^x$ at the number $a = 0$.

   Answer: $x + x^2 + \frac{x^3}{2} + \frac{x^4}{6}$ medium

   _____

5. Find the Taylor polynomial $T_3(x)$ for the function $f(x) = \frac{5x}{2+4x}$ at the point $x_0 = 0$.

   Answer: $T_3(x) = \frac{15}{48}x^3 - \frac{5}{16}x^2 + \frac{5}{4}x$ medium

   _____

6.   Use a Taylor polynomial of degree 5 to approximate the function $f(x) = \sin x$.

Answer: $x - \frac{1}{6}x^3 + \frac{1}{120}x^5$ easy

_____

7.   Write the fourth degree Taylor polynomial centered about the origin for the function $f(x) = e^{-2x}$.

Answer: $T_4(x) = 1 - 2x - 2x^2 - \frac{4}{3}x^3 + \frac{2}{3}x^4$ medium

_____

8.   Find the Taylor polynomial, $T_3(x)$, for $f(x) = xe^x$.

Answer: $T_3(x) = 0 + x + x^2 + \frac{1}{2}x^3$ easy

_____

9.   Find the second degree Taylor polynomial for $f(x) = \sqrt{x}$, centered about $a = 100$. Also obtain a bound for the error in using this polynomial to approximate $\sqrt{100.1}$.

Answer: $T_2(x) = 10 + \frac{1}{20}(x - 100) - \frac{1}{8000}(x - 100)^2$; $|\text{error}| \leq 6 \times 10^{-10}$ hard

_____

10.   Find an approximation for $\sin(.1)$ accurate to 6 decimal places (the .1 is in radians).

Answer: 0.099833 medium

_____

11.   Give the 4th degree Taylor polynomial for $f(x) = \sqrt{x}$ about the point $x = 4$. Using this polynomial, approximate $\sqrt{4.2}$. Give the maximum error for this approximation.

Answer: $T_4(x) = 2 + \frac{x-4}{4} + \frac{(x-4)^2}{64} + \frac{(x-4)^3}{512} - \frac{5(x-4)^4}{16384} + R_4$; $\sqrt{4.2} \doteq 2.049390137$;

_____

        $R_4 \doteq 0.0000000171$ hard

12. Use the 3rd degree Taylor polynomial of $f(x) = \sqrt{x}$ about $x = 4$ to approximate $\sqrt{6}$. Use the remainder term to give an upper bound for the error in the above approximation.

Answer: $\sqrt{6} \doteq T_3(6) = 2 + \frac{2}{4} - \frac{2^2}{64} + \frac{2^3}{512} = 2.453125$; $|\text{error}| \leq 0.005$ hard

13. Find the third degree Taylor polynomial, with remainder, centered at $x = 1$ for $f(x) = \ln x$. Use this result to approximate $\ln(1.2)$ and estimate the accuracy of this approximation.

Answer: $T_3(x) = (x-1) - \frac{1}{2!}(x-1)^2 + \frac{2}{3!}(x-1)^3$, $R_3(x) = \frac{-(x-1)^4}{4c^4}$ where c is some

number between 1 and x; $\ln(1.2) \doteq 0.18266$ with $R_3(1.2) \leq 0.0004$.

The approximation is accurate to 3 decimal places. hard

14. Write the Taylor polynomial at 0 of degree 4 for $f(x) = \ln(1+x)$.

Answer: $x - \frac{1}{2}x^2 + \frac{1}{3}x^3 - \frac{1}{4}x^4$ medium